Universitext

Editors

F.W. Gehring
P.R. Halmos
C.C. Moore

Universitext

Editors

F.W. Gehring
P.R. Halmos
C.C. Moore

Edwin E. Moise

Introductory Problem Courses in Analysis and Topology

Springer-Verlag
New York Heidelberg Berlin

Edwin E. Moise
Department of Mathematics
Queens College
City University of New York
Flushing, New York 11367
U.S.A.

Library of Congress Cataloging in Publication Data

Moise, Edwin E.
  Introductory problem courses in analysis and topology.

  (Universitext)
  1. Mathematical analysis—Problems, exercises, etc.
2. Topology—Problems, exercises, etc.   I. Title.
QA301.M64   515′.076   82-861
                              AACR2

9 8 7 6 5 4 3 2 1

ISBN-13:978-0-387-90701-7   e-ISBN-13:978-1-4613-8183-9
DOI: 10.1007/978-1-4613-8183-9

Preface to the Teacher

In each chapter of this book, the student finds definitions, and theorems which are guaranteed to be true. The student's job is to prove the theorems. In the problems that follow, various propositions are stated as if they were true, but many of them turn out to be false. Here the student's job is to find out the truth, whatever it may be, and establish it with a proof or a counter-example.

Any adequate explanation of the workings and the advantages of problem courses would form an essay much too long for a preface. Following are some observations.

--Any student who has creative capacities deserves an opportunity to discover, use, and develop them; and this opportunity should come soon. It should not be postponed until the time comes to write a dissertation.

--In the first few weeks of a problem course, the teacher is likely to see striking evidence of the need for such a course. At the beginning, it rarely happens that a student can write grammatical sentences which say what he really means. At the end of a problem course, almost all students can do this.

--In a lecture course it is taken for granted that whatever the teacher says is right, except for occasional lapses. If at some point the logic of a proof is less than clear, then faith and authority may take over where reason left off. But students are aware that students are often wrong, and therefore they do not accept one another's proofs unless they understand the proofs completely. This establishes the critical vigilance that the student needed all along, and it sets a new standard of thoroughness of comprehension.

--Some have supposed that problem courses are advantageous only for students of real brilliancy, but my own experience over many years indicates the contrary. The time that is "wasted" while students grope their way makes the pace of a problem course very slow. (It often happens that a whole hour is spent analyzing a "proof" which turns out to be quite worthless.) This means that a competent student is able to keep track, and finds at the end that he understands the course completely. This is a valuable experience, and for many students it is new.

The choice of material, in the courses presented here, is based on two different considerations.

I believe that "baby real-variable theory" is so fundamental that it needs to be over-learned. To examine its technical apparatus, carefully, once, is not enough. This apparatus needs to be absorbed so completely that it forms, forever after, part of the student's intuition. Some of this material is exciting, but some of it is dull. Personally, I can tolerate almost any sort of spade-work, if there is good reason to do it, and if I am doing it myself; but to watch carefully while somebody else does it is a much worse hardship. I doubt that this attitude is unusual .... For these two reasons, I think that a problem course is a good medium for learning the beginnings of analysis.

Elementary set-theoretic topology has a different advantage. It seems an ideal gymnasium for learning what one might call Applied Mathematical Logic. In this material, **anti**-intuitive results are remarkably prevalent. Thus the tiniest logical slip often leads the student to "prove" a proposition which is false. When this happens, it is plain that logical rigor is not something imposed on the student by academic authority; it is an objective necessity.

Both these courses are short, and so any student who really wants to know something about analysis and topology is going to have to take other courses in order to do so. This was intentional. I believe that every undergraduate should take at least one problem course, and probably two; but this style of study should not be unduly prolonged. If it is, it may lead to consequences which are both unnecessary and sad: the student may come to feel that nothing is worth knowing unless you discover it for yourself, so that study is an activity inappropriate for a thinker. Being aware that this state of mind is a handicap, I prefer not to **propagate it.**

Edwin E. Moise

New York, November, 1981

Table of Contents

Analysis

Chapter 1: Notations

We shall use the following standard notations of logic and set theory.

$A \subset B$	A is a subset of B.
$B \supset A$	The set B contains the set A.
$x \in A$	x is an element of the set A.
$\emptyset$	The empty set
$A \cup B$	The union of the sets A and B.
$\{x \mid (\dots)\}$	The set of all objects x such that $(\dots)$ holds.

Thus $A \cup B = \{x \mid x \in A \text{ or } x \in B\}$, and $A \cap B = \{x \mid x \in A \text{ and } x \in B\}$. If A and B have no element in common, then $A \cap B = \emptyset$.

The symbol "$\Longrightarrow$" stands for <u>implies</u>. Thus "P $\Longrightarrow$ Q" means that "if P, then Q". The symbol "$\Longleftrightarrow$" stands for "is equivalent to". Thus "P $\Longleftrightarrow$ Q" means "P if and only if Q"; that is, both P $\Longrightarrow$ Q and Q $\Longrightarrow$ P, so that the propositions P and Q are either both true or both false.

The following symbols will seldom be used in the text, but are convenient abbreviations for use in note-books and on blackboards.

$\forall$ stands for <u>always</u>, that is, for all values of the variables under discussion. For example, in the real number system, $(x+y)^2 = x^2 + 2xy + y^2$ $\forall$.

$\ni$ stands for <u>such that</u>.

$\exists$ stands for <u>there exists</u>.

$\nexists$ stands for <u>there does not exist</u>.

$\neg$ stands for <u>it is false that</u>.

For example, $\neg$ 2 + 2 = 7.

Note that when we write $A \subset B$, we allow the possibility that $A = B$. Thus $\subset$ is like $\leq$, not like $<$.

Chapter 2: The Real Numbers, Regarded as an Ordered Field

We shall suppose that the set $\mathbb{R}$ of all real numbers is given, with its binary operations $+$ and $\cdot$. The system $[\mathbb{R}, +, \cdot]$ forms a <u>field</u>. By this we mean that the following conditions hold.

<u>Closure</u>. For each x, $y \in \mathbb{R}$, $x+y \in \mathbb{R}$ and $xy \in \mathbb{R}$.

<u>The CAD Laws</u>. (Commutative, Associative, and Distributive.) For each x, y, z in $\mathbb{R}$, we have $x+y = y+x$, $xy = yx$, $(x+y) + z = x + (y+z)$, and $x(y+z) = xy + xz$.

<u>Existence of</u> 0. There is one and only one number 0 such that $x + 0 = x$ for every x.

<u>Existence of Opposites</u>. For each x there is one and only one number $-x$ such that $x + (-x) = 0$.

<u>Existence of 1</u>. There is one and only one number 1 such that $1 \cdot x = x$ for every x. And $1 \neq 0$.

<u>Existence of Reciprocals</u>. For each $x \neq 0$ there is one and only one number $1/x$ such that $x(1/x) = 1$.

We shall regard these statements as postulates; that is, we shall accept them without proving them. And since this is a course in analysis rather than the foundations of analysis, we shall also use the familiar computational procedures based on them. We also know that $xy = 0 \Longrightarrow x = 0$ or $y = 0$.

The <u>product</u> $A \times B$ of two sets A and B is the set of all ordered pairs (a,b), where $a \in A$ and $b \in B$. A <u>relation</u> on a set A is a subset R of $A \times A$. If $(a,b) \in R$, then we write aRb.

Let $<$ be a relation on a set A. Suppose that the following hold:

0.1 (Anti-reflexivity.) $a < a$ <u>never</u>.

0.2 (Transitivity.) $a < b$ and $b < c \Longrightarrow a < c$.

Then $<$ is an <u>ordering</u> of A, and the pair $[A, <]$ is an <u>ordered set</u>. Suppose also that:

0.3 (Trichotomy.) For each a, $b \in A$, one and only one of the following holds:

$$a < b, \text{ or } a = b, \text{ or } b < a .$$

Then $<$ is a <u>linear ordering</u> of A, and the pair $[A, <]$ is a <u>linearly ordered set</u>. If $a < b$, then we may also write $b > a$.

In the real number system, we have a linear ordering $<$ of $\mathbb{R}$, satisfying the following conditions.

AO. $a < b \Longrightarrow a+c < b+c$.

MO. $a > 0$ and $b > 0 \Longrightarrow ab > 0$.

Thus the real number system is a quadruplet $[\mathbb{R}, +, \cdot, <]$. Any algebraic system satisfying the postulates that we have stated so far is called

2. The Real Numbers, Regarded as an Ordered Field

an ordered field. Note that an ordered field is not merely a field which is arranged in an order: < must be related to addition and multiplication in the way described by AO and MO. But from these postulates we can derive all the elementary facts about inequalities, as we shall see.

Later we shall state a final postulate which will say, in effect, that the real number system "has no holes in it". From this it will follow that every positive number has roots of all orders. Meanwhile we shall assume that every positive number a has a positive square root. Obviously, if $x^2 = a$, then $(-x)^2 = a$. Thus every positive number a has two square roots. The positive square root of a is denoted by $\sqrt{a}$. We also define $\sqrt{0} = 0$.

The following laws of inequalities can be proved in the stated order.

Theorem 1. $a > 0 \implies -a < 0$.

Theorem 2. $a < 0 \implies -a > 0$.

Theorem 3. $1 > 0$. (This may look like a joke, but it isn't. It is supposed to be proved on the basis of the postulates that we have so far written down. We need to show that in any ordered field, the element which leaves things unchanged under multiplication is greater than the element which leaves things unchanged under addition.)

Theorem 4. $a < b$ and $c < d \implies a+c < b+d$.

Theorem 5. $a < b \implies b-a > 0$.

Theorem 6. $b-a > 0 \implies a < b$.

Theorem 7. $a < b \iff b-a > 0$.

Theorem 8. $a > 0 \implies 1/a > 0$.

Theorem 9. $a < b$ and $c > 0 \implies ac < bc$.

Theorem 10. $ac < bc$ and $c > 0 \implies a < b$.

Theorem 11. For each $c > 0$, $a < b \iff ac < bc$.

Theorem 12. For each $c < 0$, $a < b \implies ac > bc$.

Theorem 13. $c < 0 \implies 1/c < 0$.

Theorem 14. For each $c < 0$, $a < b \implies ac > bc$.

Definition. For each x,

$$|x| = \begin{cases} x & \text{if } x \geq 0 \\ -x & \text{if } x < 0. \end{cases}$$

Theorem 15. $|x| \geq 0 \ \forall$.

Theorem 16. $x^2 = |x|^2 \ \forall$.

Theorem 17. $|x| = \sqrt{x^2} \ \forall$.

Theorem 18. $|-x| = |x| \ \forall$.

Theorem 19. $|xy| = |x| \cdot |y| \ \forall$.

Theorem 20. $x \leq |x| + |y| \ \forall$.

Theorem 21. $|x+y| \leq |x| + |y| \ \forall$.

Theorem 22. For each a, b > 0, a < b $\Longrightarrow$ a^2 < b^2.

Theorem 23. For each d > 0, |x| < d $\Longleftrightarrow$ −d < x < d.

Theorem 24. For each d > 0, |x−a| < d $\Longleftrightarrow$ a−d < x < a+d.

Theorem 25. x < y $\Longrightarrow$ x < (x+y)/2 < y.

Definitions. In $\mathbb{R}$,

    (a) [a,b] = {x|a $\leq$ x $\leq$ b}.

    (b) (a,b) = {x|a < x < b}.

    (c) [a,b) = {x|a $\leq$ x < b}.

    (d) (a,b] = {x|a < x $\leq$ b}.

    (e) (−∞,a] = {x|x $\leq$ a}.

Similarly for (f) (−∞,a), (g) [a,∞), and (h) (a,∞), and (i) (−∞,∞) = $\mathbb{R}$. All such sets are called intervals. Types (a), (e), (g), (h), and (i) are called closed intervals; types (b), (f), (h), and (i) are open: and (c) and (d) are half-open. The points a and b are the end-points in cases (a)-(d); and in cases (e)-(h), a is the (only) end-point. Note that we are not regarding "infinity" or "minus infinity" as numbers; we have assigned meanings to such expressions as "[a,∞)", but in this book, at least, the symbol "∞" has no meaning at all when it stand alone.

If I is an interval, and x is a point but not an end-point of I, then x is an interior point of I.

Theorem 26. a < c $\leq$ d < e $\Longrightarrow$ d−c < e−a.

Theorem 27. a $\leq$ c $\leq$ d $\leq$ e $\Longrightarrow$ d−c $\leq$ e−a.

Theorem 28. x, y ε (a,e) $\Longrightarrow$ |x−y| < e−a.

Theorem 29. x, y ε [a,e] $\Longrightarrow$ |x−y| $\leq$ e−a.

Problem Set 2

1. In a coordinate plane, let A be the union of the lines x = 1 and x = 2. Is it possible to define a linear ordering of A?

Hereafter, we shall denote a coordinate plane by $\mathbb{R}^2$. Here $\mathbb{R}^2$ means $\mathbb{R} \times \mathbb{R}$: a coordinate plane is the product of the real numbers with themselves.

2. Is it possible to define a linear ordering of $\mathbb{R}^2$?

3. Same question, for the set $\mathbb{C}$ of all complex numbers.

4. In Theorem 22, suppose that we omit the hypothesis that a, b > 0. Is the resulting statement true?

5. In Theorem 23, suppose that we omit the hypothesis that d > 0. Is there any harm done?

6. Let R = {0,1}, and define + and · as follows:
0 + 0 = 0, 0 + 1 = 1 + 0 = 1, 1 + 1 = 0, 1·1 = 1, 1·0 = 0·1 = 0, 0·0 = 0.
This gives a system [R, +, ·]. Verify that this system is a field.

8. Is it true that every ordered field is infinite?

9. For every x and y, |x| − |y| $\leq$ |x−y|.

4

2. The Real Numbers, Regarded as an Ordered Field

    10. Is it possible to define a linear ordering < of ℂ, in such a way that
[ℂ, +, ·, <] is an ordered field?

Chapter 3: Functions, Limits, and Continuity

   Definition. Let A and B be non-empty sets, and let f be a collection of ordered pairs (x,y), such that

   (1) If $(x,y) \in f$, then $x \in A$ and $y \in B$, and

   (2) Each x in A is the first term of one and only one pair $(x,y) \in f$.

Then f is a <u>function</u>, of A into B, and we write f: A $\longrightarrow$ B .

   A is the <u>domain</u> of the function, and B is the <u>codomain</u>. For each $x \in A$, f(x) denotes the second term of the pair in f whose first term is x. If f(x) = y, then we may also write $x \longmapsto y$. For each $A' \subset A$, we define

$$f(A') = \{y | y = f(x) \text{ for some } x \subset A'\} .$$

The set f(A') is called the <u>image of</u> A'. The total image f(A) is called simply the <u>image</u>. If $x \in A$, then we say that f is <u>defined at</u> x.

   Eventually we shall need the full generality of this definition, in which A and B are any non-empty sets whatever. But for quite a while, A and B will be subsets of ℝ. The simplest case is that of a function f:A $\longrightarrow$ B, where A and B are intervals of some kind. (We allow, for example, the interval ℝ = (-∞,∞).) But even at the beginning, in discussing limits, we need to discuss functions whose domains are not intervals. Given a function f: I $\longrightarrow$ B, where I and B are intervals, and given a point <u>a</u> of I, we define the derivative of f at <u>a</u> in the usual way, as

$$f'(a) = \lim_{x \to a} \frac{f(x) - f(a)}{x-a}$$

(if such a limit exists). Thus

$$f'(a) = \lim_{x \to a} \Phi(x),$$

where

$$\Phi(x) = \frac{f(x) - f(a)}{x-a} .$$

Here the domain A of the function Φ is the <u>punctured</u> <u>interval</u> I - a, that is, the set of all points of I except for <u>a</u>. We cannot use the value x = a, because it gives a zero denominator.) Since we are going to use limits to define derivatives, we need to define limits for functions whose domains may be either intervals or punctured intervals. Hence the conditions of the following definition.

   Definition. Let I be an interval in ℝ, let $a \in I$, and let A be either I or the punctured interval I - a. Let f be a function A $\longrightarrow$ ℝ, and let $L \in$ ℝ. Suppose that for every $\varepsilon > 0$ there is a $\delta > 0$ such that for each $x \in A$,

$$0 < |x - a| < \delta \implies |f(x) - L| < \varepsilon.$$

Then

$$\text{Lim}_{x \to a} f(x) = L \; .$$

Note that in this definition, we do not care whether $f(a)$ is defined, or what its value is; the possibility $x = a$ is reuled out by the condition $0 < |x-a|$.

Note also that this definition applies in several ways, according to the choice of I and a. Thus if I is an open interval (c,d), and $a \in I$, then $\text{Lim}_{x \to a} f(x)$ is the usual "two-sided limit". If I is a closed interval $[c,d]$, then for $a = c$ we get a "right-handed limit" $\text{Lim}_{x \to a^+} f(x)$, and for $a = d$ we get a "left-handed limit" $\text{Lim}_{x \to a^-} f(x)$. Any theory based on the general definition takes care of all these cases.

In the following theorems it is assumed that I is an interval; $a \in I$; A is either I or $I - a$; and f is a function $A \to \underline{R}$.

Theorem 1. $\text{Lim}_{x \to a} f(x) = L \iff \text{Lim}_{x \to a} [f(x) - L] = 0.$

Theorem 2. $\text{Lim}_{x \to a} f(x) = 0$ and $\text{Lim}_{x \to a} g(x) = 0 \implies \text{Lim}_{x \to a} [f(x) + g(x)] = 0.$

Theorem 3. $\text{Lim}_{x \to a} f(x) = L$ and $\text{Lim}_{x \to a} g(x) = L' \implies \text{Lim}_{x \to a} [f(x) + g(x)] = L + L'.$

Theorem 4. Let $k \in R$. If $f(x) = k$ for each x in A, then $\text{Lim}_{x \to a} f(x) = k.$

(A function f as in Theorem 4 is called a constant function. Note that a constant function f, with $f(x) = k$ for each x, is different from the number k; f is a collection of ordered pairs, and k is not.)

Definition. Suppose that there are numbers $M > 0$ and $\delta > 0$ such that

$$x \in A \text{ and } |x - a| < \delta \implies |f(x)| \leq M.$$

Then f is locally bounded at a.

Theorem 5. If $\text{Lim}_{x \to a} f(x) = L$, then f is locally bounded at a.

Theorem 6. If f is locally bounded at a, and $\text{Lim}_{x \to a} g(x) = 0$, then $\text{Lim}_{x \to a} [f(x)g(x)] = 0.$

Theorem 7. If $\text{Lim}_{x \to a} f(x) = L$ and $\text{Lim}_{x \to a} g(x) = L'$, then $\text{Lim}_{x \to a} [f(x)g(x)] = LL'.$

Definition. Suppose that there are numbers $m > 0$ and $\delta > 0$ such that

$$x \in A \text{ and } 0 < |x - a| < \delta \implies |f(x)| \geq m.$$

Then f is locally bounded away from 0 at a.

(Note that this definition allows the possibility that $f(a)$ is defined and is $= 0$; in our first few theorems on limits, we are dealing only with numbers x which are close to $\underline{a}$ but different from $\underline{a}$.)

Theorem 8. If $\text{Lim}_{x \to a} f(x) = L \neq 0$, then f is locally bounded away from 0 at $\underline{a}$.

Theorem 9. Suppose that $g(x) \neq 0$ for each x in A, so that the reciprocal $1/g(x)$ gives a well-defined function $h = 1/g: A \rightarrow \mathbb{R}$. If $\lim_{x \to a} g(x) = L \neq 0$, then $1/g$ is locally bounded at a.

Theorem 10. Let g and L be as in Theorem 9. Then $\lim_{x \to a} 1/g(x) = 1/L$.

Theorem 11. Let g be as in Theorems 9 and 10, and suppose that $\lim_{x \to a} f(x) = L'$. Then $\lim_{x \to a} [f(x)/g(x)] = L'/L$.

In the last three theorems, the condition that $g(x) \neq 0$ for each $x \in A$ is not as restrictive as it looks. In investigating limits, we are concerned only with numbers x which are close to $\underline{a}$ and different from $\underline{a}$. Therefore we are free to reduce the domain A, by excluding the point $\underline{a}$, and excluding points far from $\underline{a}$.

Definition. Let I be an interval, let f be a function $I \rightarrow \mathbb{R}$, and let $a \in I$. If $\lim_{x \to a} f(x) = f(a)$, then f is <u>continuous at</u> a.

Theorem 12. Let f and g be functions $I \rightarrow \mathbb{R}$, and let $a \in I$. If f and g are continuous at a, then $f + g$ and fg are continuous at $\underline{a}$. If also $g(x) \neq 0$ for each $x \in I$, then f/g is continuous at a.

Definition. Let I be an interval, and let f be a function $I \rightarrow \mathbb{R}$. If f is continuous at each point of I, then f is <u>continuous</u>.

Theorem 13. Let f and g be functions $I \rightarrow \mathbb{R}$. If f and g are continuous, then $f + g$ and fg are continuous. If also $g(x) \neq 0$ for each x I, then f/g is continuous.

Theorem 14. Let I be an interval, let f be a function $I \rightarrow \mathbb{R}$, and let $a \in I$. Then (1) f is continuous at a if and only if (2) for every $\varepsilon > 0$ there is a $\delta > 0$ such that

$$x \in I \text{ and } |x - a| < \delta \implies |f(x) - f(a)| < \varepsilon .$$

Let I and J be intervals, and let $f: I \rightarrow \mathbb{R}$ and $g: J \rightarrow \mathbb{R}$ be functions. If $g(x) \in I$ for each $x \in J$, then we get a well-defined composite function $f(g)$, with $[f(g)](x) = f(g(x))$ for each $x \in J$.

Theorem 15. Let $f: I \rightarrow \mathbb{R}$ and $g: J \rightarrow \mathbb{R}$ be functions. If f and g are continuous, and $f(g)$ is well-defined, then $f(g)$ is continuous.

Problem Set 3

1. Define a function $f: [0,1] \rightarrow \mathbb{R}$ such that f is not continuous at any point.

Definition. Given $f: I \rightarrow \mathbb{R}$, where I is an interval, as usual. Suppose that there is a number k such that

$$|f(x_1) - f(x_2) \le k|x_1 - x_2| \, ,$$

for each x_1, $x_2 \in I$. Then f is <u>Lipschitzian</u>, and k is a <u>Lipschitz</u> <u>constant</u> for f.

Note that for $x_1 = x_2$, the inequality

$$|f(x_1) - f(x_2) \le k \, |x_1 - x_2|$$

takes the form $0 \le 0$, which always holds. Therefore, in defining the term <u>Lipschitzian</u>, we might equally well have used the inequality

$$\frac{|f(x_1) - f(x_2)|}{|x_1 - x_2|} \le k \qquad (x_1 \ne x_2) \quad .$$

2. <u>Theorem (?)</u>. Every continuous function is Lipschitzian.

3. <u>Theorem (?)</u>. If f is a continuous function $[a,b] \longrightarrow \mathbb{R}$, then f is Lipschitzian.

4. <u>Theorem (?)</u>. Every Lipschitzian function is continuous.

5. Define $f:[0,1] \longrightarrow \mathbb{R}$ as follows. (1) If x is irrational, then $f(x) = 0$. (2) If x is rational, $= p/q$ in lowest terms, then $f(x) = 1/q$. Question: At what points, if any, is f continuous?

6. Suppose that we take the definition of continuity, furnished by condition (2) in Theorem 14, and write it in various garbled forms, as follows.

(a) For every $\varepsilon > 0$ and every $\delta > 0$,

$$|x - a| < \delta \implies |f(x) - f(a)| < \varepsilon \, .$$

(b) There is an $\varepsilon > 0$ such that for every $\delta > 0$,

$$|x - a| < \delta \implies |f(x) - f(a)| < \varepsilon \, .$$

(c) For some $\varepsilon > 0$ there is a $\delta > 0$ such that

$$|x - a| < \delta \implies |f(x) - f(a)| < \varepsilon \, .$$

(d) There is a $\delta > 0$ such that for every $\varepsilon > 0$,

$$|x - a| < \delta \implies |f(x) - f(a)| < \varepsilon \, .$$

Each of the conditions (a), (b), (c), and (d) means <u>something</u>. What does each of them mean? Sketch the situations that they describe.

7. The sine function is defined as usual. Given $\theta > 0$, lay off a path going counter-clockwise around the unit circle in the xy-plane, starting at $(1,0)$, of total length θ, ending at the point $P_\theta = (x_\theta, y_\theta)$. Then $\sin\theta = y_\theta$, by definition. (For $\theta < 0$, we lay off our path clockwise to a total length $= |\theta|$.) Show that for each θ_o, the sine is continuous at θ_o.

<u>Definition</u>. Let I be an interval in $\mathbb{R}$, and let f be a function $I \longrightarrow \mathbb{R}$. Suppose that for every $\varepsilon > 0$ there is a $\delta > 0$ such that for each x_o, $x \in I$,

$$|x_o - x| < \delta \implies |f(x) - f(x_o)| < \epsilon \ .$$

Then f is _uniformly_ _continuous_.

    8. Question: How does the definition of uniform continuity differ from the definition of continuity which is given by condition (2) of Theorem 14?

    9. _Theorem (?)._ Every uniformly continuous function $I \to \mathbb{R}$ is continuous.

    10. _Theorem (?)._ Every continuous function $I \to \mathbb{R}$ is uniformly continuous.

    11. _Theorem (?)._ Every Lipschitzian function $f: I \to \mathbb{R}$ is uniformly continuous.

    12. Let I and J be intervals in $\mathbb{R}$, let $a \in J$, and let $f: I \to \mathbb{R}$ and $g: J \to \mathbb{R}$ be functions. Suppose that the image of g lies in the domain of f; that is, $g(x) \in I$ for each $x \in J$, so that the composite function $f(g): J \to \mathbb{R}$ is well-defined. _Theorem (?)._ If $\lim_{x \to a} g(x) = b$ and $\lim_{y \to b} f(y) = L$, then $\lim_{x \to a} f(g(x)) = L$.

    13. All the theorems of this section have been based on the definition of the statement $\lim_{x \to a} f(x) = L$. In that definition, it was required that the domain A be either an interval I or a punctured interval $I - a$. When (if ever) have we used this condition for A, in proving our theorems?

    14. $\lim_{x \to a} f(x) = L$ and $\lim_{x \to a} f(x) = L' \implies L = L'$. (_Warning_: you should check your proof very carefully, to make sure that it is not circular. The theorem asserts the fact that if a limit exists at all, then it is unique. Therefore this fact must not be used in the proof.)

    15. A familiar theorem of elementary calculus asserts that if $f:[a,b] \to \mathbb{R}$ is continuous, then f has a maximum value and a minimum value. That is, there are numbers x_1, x_2 [a,b] such that for each $x \in [a,b]$, we have $f(x_1) \le f(x) \le f(x_2)$. Can this be proved, on the basis of the postulates that we have stated so far for the real number system? (Bear in mind that all our postulates so far are satisfied by the _rational_ number system $[\mathbb{Q}, +, \cdot, <]$.)

Chapter 4: Integers. Sequences. The Induction Principle.

Evidently there are some subsets S of $\mathbb{R}$ which have the following properties.

(1) $1 \in S$.

(2) For each x in $\mathbb{R}$, $x \in S \Rightarrow x+1 \in S$.

For example, $S = \mathbb{R}$ satisfies these conditions, and so also does $S = [1,\infty)$.

<u>Definition</u>. $\mathbb{Z}^+$ is the intersection of all subsets S of $\mathbb{R}$ that satisfy conditions (1) and (2).

The elements of $\mathbb{Z}^+$ are called <u>positive integers</u>, and are ordinarily denoted by the letters i, j, k, m, n, p, q, r. If we add to $\mathbb{Z}^+$ the number 0, and the negatives $-n$ of the elements n of $\mathbb{Z}^+$, this gives the set $\mathbb{Z}$ of all integers.

When we state a new definition of an old idea, we need to make sure that our definition gives the results that we have a right to expect. Thus, if we couldn't prove the next six theorems, this would mean that there was something wrong with our definition of $\mathbb{Z}^+$.

<u>Theorem 1</u>. $1 \in \mathbb{Z}^+$. And for each n, $n \in \mathbb{Z}^+ \Rightarrow n+1 \in \mathbb{Z}^+$.

<u>Theorem 2.</u> (The Induction Principle.) Let S be a subset of $\mathbb{Z}^+$. If
(1) $1 \in S$ and (2) for each $n \in \mathbb{Z}^+$, $n \in S \Rightarrow n+1 \in S$, then (3) $S = \mathbb{Z}^+$.

<u>Theorem 3.</u> $\mathbb{Z}^+$ is closed under addition.

<u>Theorem 4.</u> $\mathbb{Z}^+$ is closed under multiplication.

Let K be any linearly ordered set. When we write
$$\text{Min } K = m,$$
we mean that (1) $m \in K$ and (2) $k \in K = m \le k$. Similarly,
$$\text{Max } K = M$$
means that $M \in K$ and $k \in K \Rightarrow k \le M$. Thus Min $[0,1] = 0$, but there is no such thing as Min $(0,1)$.

<u>Theorem 5.</u> $1 = \text{Min } \mathbb{Z}^+$.

<u>Theorem 6.</u> Let $n \in \mathbb{Z}^+$. If $n \ne 1$, then $n = m+1$, for some $m \in \mathbb{Z}^+$.

<u>Theorem 7.</u> For each n, $k \in \mathbb{Z}^+$,
$$k < n \Rightarrow k+1 \le n.$$
(That is, for positive integers, the condition $k < n < k+1$ is impossible.)

<u>Definition</u>. For each $k \in \mathbb{Z}^+$,
$$J_k = \{n \mid n \in \mathbb{Z}^+ \text{ and } n > k\}.$$

<u>Theorem 8.</u> For each $k \in \mathbb{Z}^+$, $k+1 = \text{Min } J_k$.

<u>Theorem 9.</u> (The Well-ordering Principle.) Let K be a non-empty set of positive integers. Then K has a least element Min K.

Analysis

   Definition. Let A be a set, and let f be a function $\mathbb{Z}^+ \to A$. Then f is
a sequence (of elements of A).

   Under this definition, a sequence is always infinite. To define finite
sequences, we proceed as follows.

   Definition. For each $n \in \mathbb{Z}^+$,

$$I_n = \{i \mid i \in \mathbb{Z}^+ \text{ and } 1 \le i \le n\} = \{1, 2, \ldots n\} .$$

Such a set is called a segment of the positive integers. Let A be a set. A
function of the type $f: I_n \to A$ is called a finite sequence (of elements of A).

   Although sequences and finite sequences are defined as functions, they are
seldom described in functional notation. Given $f: \mathbb{Z}^+ \to A$, with $f(i) = a_i$, we
ordinarily denote the sequence by $a_1, a_2, \ldots$. Similarly for a finite sequence
$a_1, a_2, \ldots, a_n$. In each case, the numbers a_i are called the terms of the
sequence.

   The definition of the limit of a sequence is modelled on that of the limit of a
function, as follows.

   Definition. Let $a_1, a_2, \ldots$ be a sequence of real numbers, and let $L \in \mathbb{R}$.
Suppose that for every $\varepsilon > 0$ (in $\mathbb{R}$) there is an $N \in \mathbb{Z}^+$ such that

$$n \in \mathbb{Z}^+ \text{ and } n \ge N \Longrightarrow |a_n - L| < \varepsilon .$$

Then

$$\operatorname*{Lim}_{n \to \infty} a_n = L .$$

A sequence of real numbers which has a limit is called convergent.

   A set A is called finite if it is empty, or if there is a finite sequence
$a_1, a_2, \ldots, a_n$ in which each element of A appears exactly once. In this case,
we say that the set A has n elements. We shall assume, without further discussion, the familiar facts about finite sets. For example, every subset of a finite
set is finite, and the union of any finite collection of finite sets is finite.

   Theorem 10. Every finite set A of real numbers is bounded; in fact, if
$A \subset \mathbb{R}$ and $A \ne \emptyset$, then A has a least element m and a greatest element M.

   Theorem 11. Let A and B be bounded subsets of $\mathbb{R}$. Then $A \cup B$ is bounded.

   Definition. A sequence of real numbers is bounded if its terms form a bounded
set. Thus $a_1, a_2, \ldots$ is bounded if there is a number M such that for each
i, $|a_i| \le M$.

   Theorem 12. Every convergent sequence of real numbers is bounded.

   Hereafter in this section it should be understood that all sequences mentioned
are sequences of real numbers.

   Theorem 13. $\operatorname*{Lim}_{n \to \infty} a_n = L \Longleftrightarrow \operatorname*{Lim}_{n \to \infty} (a_n - L) = 0.$

   Theorem 14. $\operatorname*{Lim}_{n \to \infty} a_n = L$ and $\operatorname*{Lim}_{n \to \infty} b_n = L' \Longrightarrow \operatorname*{Lim}_{n \to \infty} (a_n + b_n) = L + L'.$

4. Integers. Sequences. The Induction Principle.

Theorem 15. $\lim_{n \to \infty} a_n = L$ and $\lim_{n \to \infty} b_n = L' \implies \lim_{n \to \infty} a_n b_n = LL'$.

Definition. The sequence a_1, a_2, ... is bounded away from 0 if there is a number $\delta > 0$ such that $|a_n| \geq \delta$ for each n.

Theorem 16. If $a_n \neq 0$ for each n, and $\lim_{n \to \infty} a_n = L \neq 0$, then a_1, a_2, ... is bounded away from 0.

Theorem 17. If $a_n \neq 0$ for each n, and $\lim_{n \to \infty} a_n = L \neq 0$, then $\lim_{n \to \infty} 1/a_n = 1/L$.

Theorem 18. If $\lim_{n \to \infty} a_n = L$, $\lim_{n \to \infty} b_n = L' \neq 0$, and $b_n \neq 0$ for each n, then $\lim_{n \to \infty} a_n/b_n = L/L'$.

The Well-ordering Principle (Theorem 9) is for practical reasons another form of the Induction Principle; and for some purposes it is more convenient to use. Another convenient form is the following.

The Induction Principle. (Colloquial Form). Let P_1, P_2, ... be a sequence of propositions. If (1) P_1 is true, and (2) for each n, $P_n \implies P_{n+1}$, then (3) each of the propositions P_1, P_2, ... is true.

We call this the Colloquial Form because strictly speaking, propositions are not mathematical objects. To see that the Colloquial Form is correct, let

$$S = \{n \mid n \in \mathbb{Z}^+ \text{ and } P_n \text{ is true}\}.$$

Then (1') $1 \in S$, because P_1 is true. And (2') for each n, $n \in S \implies n+1 \in S$, because for each n, $P_n \implies P_{n+1}$. Therefore $S = \mathbb{Z}^+$, by Theorem 1, and so all the P_n's are true.

The notation a_1, a_2, ... may seem awkwardly long. If you feel the need of a shorter notation, use the notation (a_n). More explicitly, we may write

$$a_1, a_2, \ldots = (a_n) \ (n \in \mathbb{Z}^+);$$

and we may denote the finite sequence a_1, a_2, ..., a_n by

$$(a_i) \ (1 \leq i \leq n) \ .$$

It is important not to use the symbol $\{a_n\}$ as a short notation for a_1, a_2, ... The reason is that the symbol $\{a_n\}$ already has a meaning; it means the set

$$\{a_n\} = \{a_1, a_2, \ldots \} = \{x \mid x = a_i \text{ for some } i \in \mathbb{Z}^+ \} \ .$$

The set and the sequence are different, and the difference is important. For example, the sequence

$$1, -1, 1, -1, \ldots \ ,$$

where $a_i = (-1)^{i+1}$ for each i, is an infinite sequence, but it determines a finite set, namely,

$$\{1, -1, 1, -1, \ldots \} = \{1, -1\} \ .$$

13

Analysis

The sequence

$$-1, 1, -1, 1, \ldots ,$$

where $b_i = (-1)^i$ for each i, is different from (a_i), but it determines the same finite set $\{-1, -1\}$.

Chapter 5: The Continuity of IR

We have seen that the beginning of the theory of limits, of functions and of sequences, as presented in the last two sections, does not require any assumption that IR "has no holes in it"; if, instead of IR, we deal with the rational number system [Q, +, ·, <], then the theorems in Sections 3 and 4 still hold, and their proofs are exactly the same, because the rational numbers form an ordered field. On this basis, the familiar derivations of the elementary differentiation formulas become complete proofs; all that was needed, to justify them, was a valid definition of a limit, and simple theorems based on it.

To go much further, however, we need to use the fact that the real number system is <u>complete</u>, in the sense that we shall now explain.

Let A be a non-empty set of real numbers. Suppose that there is a number b such that $x \leq b$ for each $x \in A$. (Thus $A \subset (-\infty, b]$.) Then A is <u>bounded</u> <u>above</u>, and b is an <u>upper</u> <u>bound</u> of A. Similarly for <u>bounded</u> <u>below</u> and <u>lower</u> <u>bound</u>. If A is both bounded above and bounded below, then A is <u>bounded</u>. If so, there is a number M such that $|x| \leq M$ for each $x \in A$.

<u>The Completeness Postulate (CP)</u>. Let A be a non-empty set of real numbers. If A is bounded above, then A has a least upper bound. If so, the least upper bound is denoted by lub A.

<u>Theorem 1</u>. Let A be a non-empty set of real numbers. If A is bounded below, then A has a greatest lower bound.

The greatest lower bound is denoted by glb A.

<u>Theorem 2</u>. Suppose that IR is expressed as the union of two non-empty sets A and B, such that if $a \in A$ and $b \in B$, then $a < b$. Then either (1) A has a greatest element or (2) B has a least element.

The statement made in Theorem 2 is of historical interest; it was used by Richard Dedekind as a formulation of the continuity of IR. CP and Theorem 1, however, are easier to use.

<u>Theorem 3</u>. (The Archimedean Property.) For every $\epsilon > 0$ and $M > 0$ there is a positive integer n such that $n \epsilon > M$.

<u>Theorem 4</u>. $\displaystyle\lim_{n \to \infty} 1/n = 0$.

A sequence $a_1, a_2, \ldots$ of real number is <u>increasing</u> if for each n, $a_n \leq a_{n+1}$. If $a_n \geq a_{n+1}$ for each n, then the sequence is <u>decreasing</u>. (If $a_n < a_{n+1}$, then the sequence is <u>strictly</u> <u>increasing</u>. Similarly for <u>strictly</u> <u>decreasing</u>. If the set $\{a_1, a_2, \ldots\}$ of all terms of the sequence is bounded above, then the sequence is bounded <u>above</u>. Similarly for <u>bounded</u> <u>below</u>.

   Theorem 5. If a sequence of real numbers is increasing, and is bounded above, then it is convergent.

   Theorem 6. If a sequence of real numbers is decreasing, and is bounded below, then it is convergent.

   Theorem 7. If $0 \leq r < 1$, then $\text{Lim } r^n = 0$.

   Thus, in particular, $\underset{n\to\infty}{\text{Lim}} (1/2)^n = 0$.

   Let $[a_1, b_1]$, $[a_2, b_2]$, ... be a sequence of closed intervals. If for each n we have $[a_{n+1}, b_{n+1}] \subset [a_n, b_n]$, then the sequence is <u>nested</u>. Let A_1, A_2, ... be a sequence of sets. Then the intersection of all the sets A_n is denoted by

$$\bigcap_{n=1}^{\infty} A_n .$$

Thus

$$\bigcap_{n=1}^{\infty} A_n = \{a \mid a \in A_n \text{ for each } n \in \mathbb{Z}^+\} .$$

Similarly for

$$\bigcup_{n=1}^{\infty} A_n = \{a \mid a \in A_n \text{ for some } n \in \mathbb{Z}^+\} .$$

   Theorem 8. (The Nested Interval Theorem, NIP.) Let $[a_1, b_1]$, $[a_2, b_2]$, ... be a nested sequence of closed intervals in $\mathbb{R}$. Then

$$\bigcap_{n=1}^{\infty} [a_n, b_n] \neq \emptyset .$$

   The name of this theorem is abbreviated as NIP because in some treatments of $\mathbb{R}$ it is taken as a postulate.

   Theorem 9. In Theorem 8, if $\underset{i\to\infty}{\text{Lim}} (b_i - a_i) = 0$, then $\bigcap_{i=1}^{\infty} [a_i, b_i]$ is a single point $\underline{a}$.

   Definition. Let G be a collection of sets. Then G* is the union of all the elements of G. If M is a set, and $M \subset G^*$, then we say that G <u>covers</u> M.

   Theorem 10. (The Heine-Borel Theorem.) Let $[a,b]$ be a closed interval in $\mathbb{R}$, let G be a collection of open intervals, and suppose that G covers $[a,b]$. Then some finite subcollection of G covers $[a,b]$.

   Definition. Given $M \subset \mathbb{R}$, $a \in \mathbb{R}$. Suppose that every open interval containing $\underline{a}$ contains a point x of M, with $x \neq a$. Then $\underline{a}$ is a <u>limit-point</u> of M.

   Theorem 11. (The Bolzano-Weierstrass Theorem.) Every bounded infinite subset of $\mathbb{R}$ has a limit-point.

   Definition. Let A be a subset of $\mathbb{R}$, and let f be a function $A \longrightarrow \mathbb{R}^+$. If the image $f(A) = \{y \mid y = f(x) \text{ for some } x \in A\}$ <u>is bounded above</u> (or <u>bounded</u>

5. The Continuity of $\mathbb{R}$.

below, or bounded), then f is bounded above (or bounded below, or bounded).

Theorem 12. Let [a,b] be a closed interval in $\mathbb{R}$, and let f be a function [a,b] $\longrightarrow \mathbb{R}$. If f is locally bounded at each point of [a,b], then f is bounded.

Theorem 13. Let [a,b] be a closed interval in $\mathbb{R}$, and let f be continuous function [a,b] $\longrightarrow \mathbb{R}$. Then f is bounded.

Theorem 14. Let f be as in Theorem 13. Then there are numbers x_o, $x_1 \in [a,b]$ such that

(1) For each $x \in [a,b]$, $f(x) \leq f(x_o)$, and

(2) For each $x \in [a,b]$, $f(x) \geq f(x_1)$.

Under these conditions, we say that f has a maximum at x_o and a minimum at x_1.

The derivative of a function f: I $\longrightarrow \mathbb{R}$ is defined as usual. If $a \in I$, and f'(a) is defined, then f is differentiable at a. If $A \in I$, and f'(x) is defined for each $x \in A$, then f is differentiable on A. It is quite easy to show that if f is differentiable at a, then f is continuous at a.

Theorem 15. (Rolle's Theorem.) Let f be a continuous function [a,b] $\longrightarrow \mathbb{R}$, such that f is differentiable on (a,b), and suppose that f(a) = f(b) = 0. Then there is a point x_o of (a,b) such that $f'(x_o) = 0$.

Theorem 16. (The Mean-value Theorem, MVT.) Let f be a function [a,b] $\longrightarrow \mathbb{R}$, such that f is continuous on [a,b] and differentiable on (a,b). Then there is a point x_o of (a,b) such that

$$f'(x_o) = \frac{f(b) - b(a)}{b - a} \quad .$$

We shall now give a proof of the Chain Rule, because most of the "proofs" that get printed are defective.

Theorem 17. (The Chain Rule.) Let f: I $\longrightarrow \mathbb{R}$ and g: J $\longrightarrow \mathbb{R}$ be fucntions, such that the composite function h(x) = f(g(x)) is well-defined. If g is differentiable at a, and f is differentiable at g(a), then h = f(g) is differentiable at a, and

$$h'(a) = f'(g(a))g'(a) \quad .$$

Thus, if f and g are differentiable, then h is differentiable, and

$$h' = f'(g)g' \quad .$$

Proof. Let

$$\Delta g = g(a + \Delta x) - g(a) \quad ,$$

so that

$$g'(a) = \underset{\Delta x \to 0}{Lim} \frac{\Delta g}{\Delta x} \quad .$$

Let $u_o = g(a)$, and let

Analysis

$$\Delta f = f(u_0 + \Delta u) - f(u_0) \ ,$$

so that

$$f'(u_0) = \lim_{\Delta u \to 0} \Delta f / \Delta u \ .$$

Then

$$\lim_{\Delta u \to 0} [\Delta f / \Delta u - f'(u_0)] = 0 \ .$$

Define the function $\varepsilon = \varepsilon(\Delta u)$ be the conditions

$$\varepsilon(\Delta u) = \begin{cases} \Delta f / \Delta u - f'(u_0) & \text{for } \Delta u \neq 0 \\ 0 & \text{for } \Delta u = 0 \ . \end{cases}$$

Then

$$\Delta f = f'(u_0)\Delta u + \varepsilon(\Delta u)\Delta u \ ;$$

and this equation holds whenever $f(u_0 + \Delta u)$ is defined; it works when $\Delta u = 0$, because then it takes the form $0 = f'(u_0) \cdot 0 + 0 \cdot 0$. Now

$$h'(a) = \lim_{\Delta x \to 0} \frac{f(g(a+\Delta x)) - f(g(a))}{\Delta x}$$

$$= \lim_{\Delta x \to 0} (1/\Delta x)[f(u_0 + \Delta u) - f(u_0)] \ ,$$

where $\Delta u = g(a + \Delta x) - g(a) = g(a + \Delta x) - u_0$. This gives

$$h'(a) = \lim_{\Delta x \to 0} (\Delta f / \Delta x)$$

$$= \lim_{\Delta x \to 0} (1/\Delta x)[f'(u_0) + \varepsilon(\Delta u)\Delta u]$$

$$= \lim_{\Delta x \to 0} [f'(u_0)(\Delta u / \Delta x) + \varepsilon(\Delta u)(\Delta u / \Delta x)] \ .$$

Now as $\Delta x \to 0$, $\Delta u / \Delta x \to g'(a)$; $\Delta u \to 0$, and so $\varepsilon(\Delta u) \to 0$. Therefore

$$h'(a) = f'(g(a))g'(a) + 0 \ .$$

The defective proofs of this theorem use the assumption that $\Delta g \neq 0$ when Δx is sufficiently small. This condition holds whenever $g'(a) \neq 0$, but when $g'(a) = 0$ it may fail to hold. See one of the problems below.

Theorem 18. (The Betweenness Theorem.) Let f be a continuous function $[a,b] \to \mathbb{R}$. If $f(a) < y_0 < f(b) < y_0 < f(a))$, then there is an x_0 (a,b) such that $f(x_0) = y_0$.

Theorem 19. Let f be a continuous function $[a,b] \to \mathbb{R}$. Then the image $f([a,b])$ is a closed interval.

(This theorem sums up quite a lot.)

18

Theorem 20. Let $f: [a,b] \longrightarrow \mathbb{R}$ be a continuous function, and let $y_o \in \mathbb{R}$. Suppose that $f(x_o) = y_o$ for some x_o. Then there is a least such x_o.

Theorem 21. (The Squeeze Principle.) Let I be an interval, let $a \in I$, and let f, g, and h be functions which are defined at every point of I except perhaps the point a. Suppose that for each $x \in I$, with $x \neq a$, we have either $f(x) \leq g(x) \leq h(x)$ or $h(x) \leq g(x) \leq f(x)$. If $\lim\limits_{x \to a} f(x) = \lim\limits_{x \to a} h(x) = L$, then $\lim\limits_{x \to a} g(x) = L$.

Now that we have defined the term _limit-point_, we can give a more general definition of the statement $\lim\limits_{x \to a} f(x) = L$, as follows.

Definition. Let A be a non-empty set of real numbers, and let f be a function $A \longrightarrow \mathbb{R}$. Let $\underline{a}$ be a limit-point of A. Suppose that for every $\epsilon > 0$ there is a $\delta > 0$ such that

$$x \in A \quad \text{and} \quad 0 < |x-a| < \delta \implies |f(x)-L| < \epsilon \ .$$

Then $\lim\limits_{x \to a} f(x) = L$.

Under this more general definition, all the Theorems in Section 3 still hold; they did not depend on any hypothesis at all for the domain (except, of course that the domain of a function is never empty.) And the condition that $\underline{a}$ be a limit-point of A is exactly what is needed, to make the theorem stated in Problem 3.14 still true.

Problem Set 5

1. In any linearly ordered set [S,<], upper bounds and lower bounds are defined in the same way as in [$\mathbb{R}$,<]. Suppose that every non-empty subset of S which is bounded above has a least upper bound. Then [S,<] is complete.

Theorem (?). Suppose that [S,<] is complete, and let M be a non-empty set of S. If M is bounded below, then M has a greatest lower bound.

2. Show that if Theorem 2 is taken as a postulate, then CP can be proved as a theorem.

3. We define a function $f: \mathbb{R} \longrightarrow \mathbb{R}$ be the conditions (1) $f(x) = x\sin(1/x)$ for $x \neq 0$ and (2) $f(0) = 0$. Show that f is continuous everywhere, but is not differentiable at 0.

4. Let $g(x) = x^2\sin(1/x)$ for $x \neq 0$, and let $g(0) = 0$. Show that g is differentiable, and that g' is not continuous at 0.

5. You have seen that a derivative need not be continuous. Show, however, that every derivative has the betweenness property described by Theorem 18.

6. Let $S = [0,1]^2 = [0,1] \times [0,1]$. For (a,b), $(c,d) \in S$, define $(a,b) < (c,d)$ to mean that either (1) $a < c$ or (2) $a = c$ and $b < d$. Show that < is a

linear ordering of S. Is [S,<] complete, in the sense defined in Problem 1?

7. Let [R, +, ·, <] be an ordered field, not necessarily satisfying CP. In such a system, we can define the set Z^+ of positive integers, in exactly the same way as at the beginning of this section. Suppose we know that in R,

$\lim_{n \to \infty} (1/n) = 0$. Show that the system has the Archimedean Property described in Theorem 3. (Thus this innocent-looking limit relation has unsuspected depth. Some ordered fields are not Archimedean.)

8. For every $\alpha < \beta$ in R there is a rational number p/q such that $\alpha < p/q < \beta$.

9. Every positive real number has roots of all order. That is, for each a > 0 in R, and each $n \in Z^+$, there is an x > 0 in R such that $x^n = a$.

10. Consider the equation

$$\frac{1-x}{x} = x .$$

By elementary methods, we can show that this equation has a positive root, namely, $x = (-1 + \sqrt{5})/2$. Show that x cannot be rational.

11. Show that every open interval in R contains an irrational number.

12. Let a_1, a_2, ... be a sequence of real numbers. Suppose that for every $\varepsilon > 0$ there is an $N \in Z^+$ such that

$$m, n \in Z^+ \text{ and } m, n \geq N \Rightarrow |a_n - a_m| < \varepsilon .$$

Then a_1, a_2, ... is called a _regular_ sequence, or a Cauchy sequence.

Theorem. Every convergent sequence is regular. Query: Does the proof of this theorem require the use of CP?

13. Theorem. Every regular sequence is convergent. (Same query as in Problem 12.)

Definition. Let n_1, n_2, ... be a strictly increasing sequence of positive integers. Let a_1, a_2, ... be a sequence; and for each i, let $b_i = a_{n_i}$. Then b_1, b_2, ... is a _subsequence_ of a_1, a_2, ...

14. Theorem. Let a_1, a_2, ... be a bounded sequence of real numbers. Then a_1, a_2, ... has a convergent subsequence.

15. Theorem (?). Let [a,b] be a closed interval in R, and let G be a collection of open intervals, covering [a,b]. Then there is a number $\delta > 0$ such that for each $x \in [a,b]$ there is an element g_x of G such that

$$(x - \delta, x + \delta) \subset g_x .$$

Note that by the Heine-Borel Theorem, this proposition reduces immediately to a case in which G is a finite collection.

16. Let f be a differentiable function (a,b) $\rightarrow$ R, and let $x_0 \in (a,b)$. Show that for every $\delta > 0$ such that

20

$$a < x_o - \delta < x < x_o < x' < x_o + \delta < b$$

$$\Rightarrow \left| \frac{f(x) - f(x')}{x - x'} - f'(x_o) \right| < \epsilon .$$

Briefly,

$$\underset{\substack{x \to x_o^- \\ x' \to x_o^+}}{\text{Lim}} \frac{f(x) - f(x')}{x - x'} = f'(x_o) .$$

17. Show that if the Bolzano–Weirstrass Theorem (BWT) is taken as a postulate, in place of CP, then all of the following can be proved: (a) Theorem 3. (b) Theorem 4. (c) Theorem 8. (d) CP. Thus, in any ordered field, CP and BWT are equivalent; and so BWT is a complete description of the continuity of $\mathbb{R}$.

18. Let A and B be non-empty sets of real numbers, with $A \subset B$. Then

   (a) If B is bounded above, then lub $A \le$ lub B;

and

   (b) If B is bounded below, then glb $B \le$ glb A; and therefore

   (c) If B is bounded, then glb $B \le$ glb $A \le$ lub $A \le$ lub B.

19. Let f and g be differentiable functions $[a,b] \to \mathbb{R}$, such that f' and g' never vanish simultaneously. (That is, $f'(t)^2 + g'(t)^2 > 0$ for each point t of $[a,b]$.) If $g(a) \ne g(b)$, then there is a point t_o, between a and b, such that

$$\frac{f(b) - f(a)}{g(b) - g(a)} = \frac{f'(t_o)}{g'(t_o)} .$$

If $f(a) \ne f(b)$, then there is a point t_o such that the recriprocals of those fractions are equal.

(This has a geometric meaning. The functions f and g define a plane path $p:[a,b] \to \mathbb{R}^2$, with $p(t) = (g(t),f(t))$; and if f' and g' never vanish simultaneously, then at each point the path has a well-defined tangent. The theorem asserts that if the end-points $p(a)$ and $p(b)$ are different, then there is a point $p(t_o)$ where the tangent is parallel to the line through the end-points.)

Chapter 6: The Riemann Integral of a Bounded Function

Let $[a,b]$ be a closed interval in $\mathbb{R}$, let f be a bounded function $[a,b] \to \mathbb{R}$, and let M be a bound for f, so that $|f(x)| \le M$ for each $x \in [a,b]$. The notations $[a,b]$, f, and M will be used in this sense throughout this section.

Theorem 1. For each x_1, $x_2 \in [a,b]$, $|f(x_1) - f(x_2)| \le 2M$.

A net over $[a,b]$ is a finite sequence

$$N: x_0, x_1, \ldots, x_n,$$

where

$$x_0 = a, \ x_n = b, \ x_i < x_{i+1} \quad \text{for} \quad 0 \le i < n.$$

Thus the points x_i decompose $[a,b]$ into n closed intervals of the type $[x_{i-1}, x_i]$ $(1 \le i \le n)$. For $1 \le i \le n$, let

$$\Delta x_i = x_i - x_{i-1}.$$

For each i from 1 to n, let

$$M_i = \text{lub} \ \{f(x) | x_{i-1} \le x \le x_i\},$$

$$m_i = \text{glb} \ \{f(x) | x_{i-1} \le x \le x_i\}.$$

(The indicated lub and glb exist, because the indicated sets are non-empty, and are bounded above and below respectively.). Let

$$U(N) = \sum_{i=1}^{n} M_i \Delta x_i,$$

$$L(N) = \sum_{i=1}^{n} m_i \Delta x_i.$$

$U(N)$ is called the upper sum of f over N, and $L(N)$ is called the lower sum.

Theorem 2. Let N_1 and N_2 be nets over $[a,b]$. Then

$$U(N_1) \ge L(N_2)].$$

Theorem 3. Let U be the set of all upper sums of f, over all nets N. Then U is bounded below.

Theorem 4. Let L be the set of all lower sums of f, over all nets N. Then L is bounded above.

Definitions.

$$\overline{\int_a^b} f = \text{glb} \ U, \qquad \underline{\int_a^b} f = \text{lub} \ L.$$

6. The Riemann Integral of a Bounded Function

These are called the <u>upper integral</u> and the <u>lower integral</u>, respectively, of f over [a,b].

The <u>mesh</u> $|N|$ of a net N is the largest of the numbers Δx_i. When we write

$$\lim_{|N| \to 0} U(N) = k ,$$

this means that for every $\varepsilon > 0$ there is a $\delta > 0$ such that

$$|N| < \delta \implies |U(N) - k| < \varepsilon .$$

Similarly,

$$\lim_{|N| \to 0} L(N) = k'$$

means that for every $\varepsilon > 0$ there is a $\delta > 0$ such that

$$|N| < \delta \implies |L(N) - k'| < \varepsilon .$$

Note that here we are not dealing with limits of functions: N is not determined when $|N|$ is known, and so U(N) and L(N) are not functions of $|N|$. But the old $\varepsilon - \delta$ scheme for defining a limit relation works in the same way in the present much more general situation.

<u>Theorem 5.</u> $\displaystyle \overline{\int_a^b} f = \lim_{|N| \to 0} U(N).$

<u>Theorem 6.</u> $\displaystyle \underline{\int_a^b} f = \lim_{|N| \to 0} L(N).$

<u>Definition.</u> If $\displaystyle \overline{\int_a^b} f = \underline{\int_a^b} f$, then f is <u>integrable</u> (on [a,b], in the sense of Riemann), and the <u>integral</u> of f (over [a,b]) is

$$\int_a^b f = \overline{\int_a^b} f = \underline{\int_a^b} f .$$

Theorem 7. If (1) f is integrable over [a,b], then (2) for every $\varepsilon > 0$ there is a net N over [a,b] such that $U(N) - L(N) < \varepsilon$.

Theorem 8. Let f be a bounded function [a,b] $\longrightarrow \mathbb{R}$. Suppose that (1) for every $\varepsilon > 0$ there is a net N over [a,b] such that $U(N) - L(N) < \varepsilon$. Then (2) f is integrable over [a,b].

Theorem 9. If f is integrable over [a,b], and $[c,d] \subset [a,b]$, then f is integrable over [c,d].

Strictly speaking, the domain of a function is part of the definition of the function; and so, if we use a smaller domain, we get a different function. In

general, if f is a function $A \longrightarrow B$, and $A' \subset A$, then $f|A'$ (the <u>restriction of</u> f
<u>to</u> A') is the function $A' \longrightarrow B$ such that for each $x \in A'$, $(f|A')(x) = f(x)$.
That is,

$$f|A' = \{(x,y)\,|\,x \in A'\ \text{and}\ (x,y) \in f\}\ .$$

Strictly speaking, therefore, in Theorem 9 we really mean that $f|[c,d]$ is
integrable. But our slight abuse of language, in Theorem 9 and in similar cases
hereafter, is customary, and seems harmless.

   <u>Theorem 10</u>. If f is integrable over $[a,b]$ and also over $[b,c]$, then f is
integrable over $[a,c]$, and

$$\int_a^c f = \int_a^b f + \int_b^c f\ .$$

   <u>Theorem 11</u>. If f is integrable over $[a,e]$, and $a \leq b < c < d \leq e$, then

$$\int_b^d f = \int_b^c f + \int_c^d f.$$

   <u>Definitions</u>. $\displaystyle\int_a^a f = 0.$ If f is integrable over $[a,b]$, then

$$\int_b^a f = -\int_a^b f\ .$$

   <u>Theorem 12</u>. Suppose that f is integrable over $[a,e]$, and let
b, c, d $\in [a,e]$. Then

$$\int_b^d f = \int_b^c f + \int_c^d f\ .$$

   Note that this theorem applies in a total of 13 cases.

   Let $f: [a,b] \longrightarrow \mathbb{R}$ be bounded, as before; let $N = x_0,\ x_1,\ \ldots,\ x_n$ be a net
over $[a,b]$; and let $S = x'_1,\ x'_2,\ \ldots x'_n$ be a finite sequence, such that for each
i, $x_{i-1} \leq x'_i \leq x_i$. Then S is a <u>sample</u> of N, and the sum

$$\textstyle\sum(N,S) = \sum_{i=1}^n f(x'_i)\Delta x_i$$

is called a <u>sample</u> <u>sum</u> <u>of</u> f <u>over</u> N. When we write

$$\text{Lim}\ \textstyle\sum(N,S) = k,$$
$$|N| \to 0$$

this means that for every $\epsilon > 0$ there is a $\delta > 0$ such that

$$|N| < \delta \implies |\ \textstyle\sum(N,S) - k| < \epsilon\ .$$

That is, $|N| \approx 0 \implies \sum(N,S) \approx k$, independently of the choice of the sample S.

6. The Riemann Integral of a Bounded Function

Theorem 13. If f is integrable, then

$$\text{Lim}_{|N| \to 0} \Sigma(N,S) = \text{Lim}_{|N| \to 0} \sum_{i=1}^{n} f(x'_i) \Delta x_i = \int_a^b f .$$

Theorem 14. Suppose that f is integrable on [a,b], and that for each $x \in [a,b]$, $m \leq f(x) \leq M$. Then

$$m(b-a) \leq \int_a^b f \leq M(b-a).$$

Theorem 15. Suppose that f is integrable on [a,b]. Let $x_0 \in [a,b]$ and let (c,d) be an open interval containing x_0, such that

$$y \in (c,d) \cap [a,b] \implies m \leq f(y) < M .$$

Then

$$x \in (c,d) \cap [a,b] \text{ and } x \neq x_0 \implies m \leq \frac{1}{x-x_0} \int_{x_0}^x f \leq M .$$

Theorem 16. Let f be integrable over [a,b]. For each $x \in [a,b]$, let $F(x) = \int_a^x f$. Then F is Lipschitzian (and therefore continuous).

(Note that this theorem does not require that f be continuous.)

Theorem 17. Let f and F be as in Theorem 16, let x_0 be a point of the open interval (a,b), and suppose that f is continuous at x_0. Then F is differentiable at x_0, and $F'(x_0) = f(x_0)$.

Proposition 18. If in Theorems 14, 15, 16, and 17 we use the upper integral or the lower integral, instead of the integral, then the resulting statements are true.

-- -- -- -- --

In this section, we have used the notation $\int_a^b f$, instead of the usual $\int_a^b f(x)dx$. The short notation is adequate, for our present purposes, because the integral is determined if the function f and the interval [a,b] (or [b,a]) are known. The usual notation is needed only for various other purposes, for example, to tell us which of the variables in an integrand is the "variable of integration". For example,

$$\int_0^1 (x^2 y + y^3) dx = \int_0^1 f ,$$

where $f(x) = x^2 y + y^3$, and y is regarded as a constant. Thus

$$\int_0^1 (x^2 y + y^3) \, dx = [x^3 y/3 + y^3 x]_{x=0}^{x=1} = y/3 + y^3;$$

while

$$\int_0^1 (x^2y + y^3)dy = [x^2y^2/2 + y^4/4]_{y=0}^{y=1} = x^2/2 + 1/4 \ .$$

In cases like this, the "differential" dx or dy is used merely to make it clear
what function we are supposed to be integrating. Differentials are also useful as a
guide to various formal processes, such as integration by substitution. But in the
present section, if we used the notation $\int_0^b f(x)dx$, then the final "(x)dx" would
clutter up our formulas to no good purpose.

Problem Set 6

1. Theorem 19. $\overline{\int_a^b} f \leq \underline{\int_a^b} f \ .$

2. Theorem 20. Let $[a,b] = [0,1]$, and for each $x \in [0,1]$ let $f(x) = x$.
Then f is integrable.

3. Theorem 21. For each $x \in [0,2]$ let $f(x) = x^2$. Then f is integrable.

4. Theorem 22. Given f: $[a,b] \longrightarrow \mathbb{R}$, as before. Suppose that f is
increasing; that is, $x_1 < x_2 \Rightarrow f(x_1) \leq f(x_2)$. Then f is integrable.

Thus, to be integrable, a function need not be continuous.

5. Consider the function defined in Problem 3.5 (that is, Problem 5 of
Problem Set 3.) Is this function integrable? Why or why not?

6. Every Lipschitzian function is integrable. (For the definition of
Lipschitzian, see Problem 3.2.)

7. Theorem 23. Suppose that f:$[a,b] \longrightarrow \mathbb{R}$ is integrable, and let F be
an anti-derivative of f, that is, a function such that F' = f. Then

$$\int_a^b f = F(b) - F(a) \ .$$

Obviously this is a variation on the Fundamental Theorem of Integral Calculus.
The proof is logically simple, but may not be easy to think of.

8. Theorem (?). Let f be a function $[a,b] \longrightarrow \mathbb{R}$, and for each positive
integer n, let f_n be a function $[a,b] \longrightarrow \mathbb{R}$. Suppose that (1) f is integrable,
(2) for each n, f_n is integrable, and (3) for each $x \in [a,b]$, $\underset{n \to \infty}{\text{Lim}} f_n(x) = f(x)$.
Then $\int_a^b f_n = \int_a^b f \ .$

Let f be a function $[a,b] \longrightarrow \mathbb{R}$. Suppose that f is differentiable, in the
usual sense, at every point of the open interval (a,b), and that at the end-points
a and b the "right-handed" and "left-handed" derivatives

$$f'_+(a) = \lim_{\Delta x \to 0^+} (1/\Delta x)[f(a+\Delta x) - f(a)],$$

$$f'_-(b) = \lim_{\Delta x \to 0^-} (1/\Delta x)[f(b+\Delta x) - f(b)]$$

are well-defined. Then we say that f is <u>differentiable</u> on $[a,b]$. Note that under this definition, if f is differentiable on $[a,b]$ and also on $[b,c]$, it does not follow that f is differentiable on $[a,c]$; the latter conclusion follows only if we also know that $f'_-(b) = f'_+(b)$.

9. <u>Theorem 24</u>. If f is differentiable on $[a,b]$, and f' is continuous on $[a,b]$, then f is integrable on $[a,b]$.

Note that the sufficient conditions for integrability that we have found so far cover quite a lot of ground. They apply to most of the continuous functions dealt with in elementary calculus, once we have verified that these functions are really continuous.

10. <u>Theorem 25</u>. For each $x \in \mathbb{R}$, let $f(x) = |x|$. Then f is integrable on every closed interval.

11. <u>Theorem 26</u>. Let f be a continuous function $[a,b] \to \mathbb{R}$. Then f is integrable.

12. Let f be a function $(a,b) \to \mathbb{R}$. If f is increasing, and is bounded below, then $\lim_{x \to a^+} f(x)$ exists, and is equal to $glb\ f((a,b))$.

13. Let f be a bounded function $I \to \mathbb{R}$, where I is an interval. For each non-empty set $A \subset I$, let

$$Osc\ (f,A) = lub\ f(A) - glb\ f(A) .$$

Then $Osc\ f(A)$ is called the <u>oscillation of</u> f <u>on</u> A.

<u>Theorem</u>. $A \subset B \subset I \implies Osc(f(A) \leq Osc(f,B)$.

Chapter 7: Necessary and Sufficient Conditions for Integrability

Theorem 1. Every continuous function $[a,b] \to \mathbb{R}$ is uniformly continuous.

There are three natural proofs of this theorem, using (a) The Nested Interval Theorem, (b) the Heine–Borel Theorem, and (c) the Bolzano–Weierstrass Theorem respectively.

Theorem 2. Every continuous function $[a,b] \to \mathbb{R}$ is integrable.

There is a "cute" proof of Theorem 2, not using uniform continuity; hence Problem 6.11. But the proof based on Theorem 1 seems conceptually more natural.

Definition. Let f be a bounded function $[a,b] \to \mathbb{R}$, and let $x_0 \in [a,b]$. For each $\delta > 0$, let

$$m(x_0, \delta) = \text{glb } \{f(x) \mid x \in [a,b] \quad \text{and} \quad |x_0 - x_0| < \delta\},$$

$$M(x_0, \delta) = \text{lub } \{f(x) \mid x \in [a,b] \quad \text{and} \quad |x - x_0| < \delta\},$$

$$\mathcal{O}(x_0, \delta) = M(x_0, \delta) - m(x_0, \delta) .$$

Theorem 3. Under the conditions of the preceding definition, $\mathcal{O}(x_0, \delta)$ approaches a limit, as $\delta \to 0^+$.

Definition. Let f be a bounded function $[a,b] \to \mathbb{R}$, and let $x_0 \in [a,b]$. Then

$$\mathcal{O}(x_0) = \underset{\delta \to 0^+}{\text{Lim}} \; \mathcal{O}(x_0, \delta).$$

$O(x_0)$ is called the _oscillation_ of f at x_0.

Definition. Let M be a subset of $\mathbb{R}$. If M contains all the limit points of M, then M is _closed_.

Note that if M has no limit points, then M is automatically closed. Thus, in particular, every finite set is closed.

Definition. Let f be a bounded function $[a,b] \to \mathbb{R}$, and let ε be a positive number. Then

$$M_\varepsilon = \{x \mid x \in [a,b] \quad \text{and} \quad \mathcal{O}(x) \geq \varepsilon\} .$$

Theorem 4. Every set M_ε is closed.

Definition. Let M be a closed bounded set of real numbers. Suppose that for every $\alpha > 0$ there is a finite collection $\{(a_i, b_i)\}$ $(1 \leq i \leq n)$ of open intervals, covering M, such that

$$\sum (b_i - a_i) < \alpha .$$

Then M is null (or of measure 0).

This definition will soon be generalized so as to apply to sets which are not necessarily closed.

Theorem 5. Let f be a bounded function $[a,b] \rightarrow \mathbb{R}$. Suppose that for each $\epsilon > 0$, M_ϵ is null. Then f is integrable.

Theorem 6. Let f be an integrable function on $[a,b]$. Then for each $\epsilon > 0$, M_ϵ is null.

Theorem 7. Let M be a closed set of real numbers. Then the complement $\mathbb{R} - M$ is the union G^* of a collection G of open intervals.

Theorem 8. Let M be a closed bounded set of real numbers, and let H be a collection of open intervals, covering M. Then some finite subcollection of H covers M.

Definition. Let A be a set. Suppose that (1) $A = \emptyset$, or (2) A is finite, or (3) there is a sequence in which each $a \in A$ appears at least once as a term. Then A is countable.

Note that Case (3) includes Case (2), since $\{a_1, a_2, \ldots, a_n\} = \{a_1, a_2, \ldots, a_n, a_n, a_n, \ldots\}$.

Departing from the logical order, for a moment, by using infinite series, we now generalize the definition of null, as follows.

Definition. Let M be a subset of $\mathbb{R}$ (not necessarily closed or bounded). Suppose that for each $\alpha > 0$ there is a countable collection $\{(a_i, b_i)\}$ of open intervals $(i = 1, 2, \ldots, n$ or $i = 1, 2, \ldots)$ such that $\{(a_i, b_i)\}$ covers M and such that

$$\sum_i (b_i - a_i) < \alpha .$$

Then M is null.

Note that the indicated sum may be finite or infinite. By Theorem 8, this definition of null agrees with our earlier and more special definition.

Theorem 9. Let M_1, M_2, ... be a sequence of closed and bounded sets of real numbers. If M_i is null for each i, then $\bigcup_{i=1}^{\infty} M_i$ is null.

Theorem 10. Let f be a bounded function $[a,b] \rightarrow \mathbb{R}$, and let M_o be the set of all points of $[a,b]$ where f is not continuous. Then (1) M_o is null if and only if (2) for each $\epsilon > 0$, M_ϵ is null.

Theorem 11. Let f be a bounded function $[a,b] \rightarrow \mathbb{R}$. Then (1) f is integrable if and only if (2) M_0 is null.

If M_0 is null, then we say that f is continuous almost everywhere.

Problem Set 7

1. In Theorem 7, can the collection G of open intervals be chosen so that every two of them are disjoint?

2. Every subset of a countable set is countable. (If you make a careful re-examination of the definition of _countable_, the chances are that you can find a very short proof of this one. Given $B \subset A$, where A is countable, the theorem reduces easily to the case in which $B \neq \emptyset$ and $A = \{a_1, a_2, \ldots\}$.)

3. No interval [a,b] is null.

4. Every countable set is null.

5. $\mathbb{R}$ is not countable.

6. Let M be a subset of $\mathbb{R}$. Let K be the set of all (finite!) numbers of the type

$$k = \sum_{i=1}^{\infty} (b_i - a_i),$$

where $\{(a,b), (a_2, b_2), \ldots\}$ is a countable collection of open intervals, covering M. If $K \neq 0$, then glb K is called the _outer measure_ of M, and is denoted by $|M|$. If $K = \emptyset$, then we write $|M| = \infty$. Thus M is null if and only if $|M| = 0$.

Theorem. For each interval $M = [a,b]$ or $M = (a,b)$, we have $|M| = b - a$.

7. Let M be the set of all irrational numbers on [0,1]. Then $|M| = 1$.

8. Prove or disprove: _Theorem (?)_. Let M be a closed bounded subset of $\mathbb{R}$. if M is not null, then M contains some interval.

9. Prove or disprove: _Theorem (?)_. Let M be a bounded subset of $\mathbb{R}$. If M is null, then M is countable.

10. Recall the function defined in Problem 3.5. Discuss the question of the integrability of f, using Theorem 11. (Compare with Problem 6.5.)

11. Let F and G be differentiable functions $[a,b] \to R$. If $F(a) = G(a)$, and $F'(x) = G'(x)$ for each $x \in (a,b)$, then $F(x) = G(x)$ for each $x \in [a,b]$.

12. _Theorem_. (The Fundamental Theorem of Integral Calculus; Classical Form.) Let F and f be continuous functions $[a,b] \to \mathbb{R}$, such that $F' = f$. Then

$$\int_a^b f = F(b) - F(a) .$$

(Compare with Problem 6.7, in which we required merely that f be integrable. The extra generality of Problem 6.7 is not needed for any purpose that I know of.)

13. For each x, let

$$F(x) = \int_0^{x^2} \sqrt{1 + t^{10}} \, dt.$$

Find $F'(x)$.

7. Necessary and Sufficient Conditions for Integrability

14. **Theorem**. Let f and g be integrable functions $[a,b] \longrightarrow \mathbb{R}$. Then $f+g$ and fg are integrable; and if $g(x) \neq 0$ for each point x of $[a,b]$, then f/g is integrable.

15. Let $f:[a,b] \longrightarrow \mathbb{R}$ be a continuous function. Let ϕ be a continuously differentiable function $[\alpha, \beta] \longrightarrow [a,b]$, such that ϕ is strictly increasing, $\phi(\alpha) = a$, and $\phi(\beta) = b$. For each net $N = t_0, t_1, \ldots, t_n$ over $[\alpha, \beta]$, let $\phi(N) = \phi(t_0), \phi(t_1), \ldots, \phi(t_n) = x_0, x_1, \ldots, x_n$, so that $\phi(N)$ is a net over $[a,b]$. Let $\Delta t_i = t_i - t_{i-1}$, $\Delta x_i = x_i - x_{i-1}$. Prove the following.

(a) There is a sample $S = t_1', t_2', \ldots, t_n'$ of N such that for each i, $\Delta x_i = \phi'(t_i')\Delta t_i$,

(b) $\displaystyle \lim_{|N| \to 0} |\phi(N)| = 0$.

(c) $\displaystyle \int_a^b f = \int_\alpha^\beta f(\phi)\phi'$.

Here (c) is of course the standard formula for integration by substitution.

16. Suppose that in Problem 15 we have given not that f and ϕ' are continuous, but merely that they are integrable. Do conclusions (a), (b), and (c) still follow? Why or why not?

17. Independently of Problem 15, show that if ϕ is a continuously differentiable function $[a,b] \longrightarrow [\phi(a), \phi(b)]$, and f is a continuous function $[\phi(a), \phi(b)] \longrightarrow \mathbb{R}$, then

$$\int_{\phi(a)}^{\phi(b)} f = \int_a^b f(\phi)\phi' \ .$$

(The point of this problem is that 15(c) is easier to verify than to derive. Note also that in Problem 17 it is not required that ϕ be increasing or decreasing. For example, we might have $(\phi)(t) = \sin t$, $a = 0$, $b = 10,000\pi + \pi/2$, $\phi(a) = 0$, $\phi(b) = 1$.)

18. **Theorem (?)**. The function $f(x) = \sqrt{x}$ is uniformly continuous on its entire domain $[0, \infty)$.

Chapter 8: Invertible Functions. Arc-length and Path-length.

Definition. Let I be an interval in $\mathbb{R}$, let f be a function $I \to \mathbb{R}$, and let $J = f(I)$. If $f(x) = f(x') \Rightarrow x = x'$, then f is invertible. If so, there is function $f^{-1}: J \to I$ such that for each $x \in I$, $f^{-1}(f(x)) = x$, and for each $y \in J$, $f(f^{-1}(y)) = y$. (To be precise, $f^{-1} = \{(y,x)|(x,y) \in f\}$.) f^{-1} is called the inverse of f.

Here the interval I may be finite or infinite, and the latter case is important. For example, in the most elegant treatment of exponentials and logarithms, due to Niels Abel, we start with the definition of

$$\ln: (0, \infty) \to \mathbb{R},$$

$$\ln x = \int_1^x dt/t.$$

Here $I = (0, \infty)$. We show that the image J is all of $\mathbb{R}$, and that $\ln$ is invertible. We then define

$$\exp = \ln^{-1}: \mathbb{R} \to (0, \infty),$$

and it turns out that $\exp x = e^x$ in the usual sense. (See Theorems 11-32 below.)

Definition. Let I be an interval, and let f be a function $I \to \mathbb{R}$. If f is either increasing (with $x < x' \Rightarrow f(x) \leq f(x')$) or decreasing (with $x < x' \Rightarrow f(x) \geq f(x')$), then f is monotonic. If f is either strictly increasing (with $x < x' \Rightarrow f(x) > f(x')$), or strictly decreasing, then f is strictly monotonic.

Theorem 1. If f is invertible and continuous, then f is strictly monotonic.

When we write

$$f: I \twoheadrightarrow J,$$

the double-headed arrow means that the image is all of J.

Theorem 2. If $f: I \twoheadrightarrow J$ is invertible and continuous, then J is an interval (finite or infinite).

Theorem 3. Let f be a continuous function $I \to \mathbb{R}$. Suppose that f is differentiable at every interior point of I, and that either (1) $f'(x) > 0$ at every interior point of I or (2) $f'(x) < 0$ at every interior point of I. Then f is invertible.

Theorem 4. If f is invertible and continuous, then f^{-1} is continuous.

Theorem 5. For each x, let $f(x) = x^2$, and let $Sq = f|[0, \infty): [0, \infty) \to \mathbb{R}$. Then Sq is invertible.

8. Invertible Functions. Arc-length and Path-length.

Definition. For each $x \geq 0$, $\sqrt{x} = Sq^{-1}x$.

- - - - - - - -

Consider now a continuous function $f: [a,b] \to \mathbb{R}$. Let $N = x_0, x_1, \ldots, x_n$ be a net over $[a,b]$. For each i, let $y_i = f(x_i)$, $P_i = (x_i, y_i) \in \mathbb{R}^2$, and for $1 \leq i \leq n$, let $P_{i-1}P_i$ be the distance between P_{i-1} and P_i, so that by the usual formula we have

$$P_{i-1}P_i = \sqrt{(x_i - x_{i-1})^2 + (y_i - y_{i-1})^2} \ .$$

Let

$$S(N) = \sum_{i=1}^{n} P_{i-1}P_i \ ,$$

so that $S(N)$ is the length of a broken line inscribed in the graph of f. If $S(N)$ approaches a limit L, as $|N| \to 0$, then the graph is <u>rectifiable</u>, and L is its <u>length</u>.

Example 1. There is a continuous function $f: [a,b] \to R$ such that the graph of f is not rectifiable.

Example 2. There is a differentiable function $[a,b] \to \mathbb{R}$ such that the graph of f is not rectifiable.

Theorem 6. Let f be a continuous function $[a,b] \to \mathbb{R}$, and suppose that f has a continuous derivative on $[a,b]$. Then the graph of f is rectifiable, and

$$L = \int_a^b \sqrt{1 + f'(x)^2} \ dx \ .$$

Theorem 7. In Theorem 6, if f' is integrable, but not necessarily continuous, then the conclusion still follows.

Consider now two continuous functions f and $g: [a,b] \to \mathbb{R}$. We define

$$p: [a,b] \to \mathbb{R}^2$$

by the condition

$$p(t) = (f(t), g(t)) \ .$$

Such a function p is called a <u>path</u> (in $\mathbb{R}^2$). Let $N = t_0, t_1, \ldots, t_n$ be a net over $[a,b]$; for each i, let

$$P_i = p(t_i) = (x_i, y_i) = (f(t_i), g(t_i)),$$

and let

$$S(N) = \sum_{i=1}^{n} P_{i-1}P_i \ .$$

If $S(N)$ approaches a limit L, as $|N| \to 0$, then p is <u>rectifiable</u>, and L is the <u>length</u> of p.

33

Theorem 8. Let p be a path, with $p(t) = (f(t), g(t))$. If f and g have continuous derivatives, then p is rectifiable, and

$$L = \int_c^b \sqrt{f'(t)^2 + g'(t)^2} \, dt .$$

Let $\phi: [c,d] \longrightarrow [a,b]$ be a continuous and strictly increasing function, and let $p: [a,b] \longrightarrow \mathbb{R}^2$ and $q: [c,d] \longrightarrow \mathbb{R}^2$ be paths, such that for each $t \in [c,d]$, $p(\phi(t)) = q(t)$. Then p and q are equivalent.

Theorem 9. Let p and q be equivalent paths in $\mathbb{R}^2$. If p is rectifiable, then q is rectifiable, and has the same length.

Following is an outline of Niels Abel's treatment of exponentials and logarithms. The reader should either recall the proofs or work them out. We start with a preliminary result, obviously important in its own right.

Theorem 10. Let f be an invertible function $I \longrightarrow J$, where I and J are intervals, and J is the image. If f is differentiable, and $f'(x) \neq 0$ for each $x \in I$, then for each $y \in J$ we have

$$f^{-1}{}'(y) = 1/f'(f^{-1}(y) .$$

(To prove this, it is not necessary (or good) to use "variables" in a mystical way. Since $f'(x)$ is never zero, it follows that for each point x_0 of I, and each $\epsilon > 0$, there is a $\delta > 0$ such that

$$x \in I \text{ and } 0 < |x - x_0| < \delta \Longrightarrow \left| \frac{x - x_0}{f(x) - f(x_0)} - \frac{1}{f'(x_0)} \right| < \epsilon .)$$

Definition. For each $x > 0$,

$$\ln x = \int_1^x \frac{dt}{t} .$$

Theorem 11. For each $x > 0$, $\ln' x = 1/x$.

Theorem 12. $\ln 1 = 0$.

Theorem 13. For each $k > 0$, let $f_k(x) = \ln kx$. Then $f'_k(x) = 1/x$.

Theorem 14. For each $k, x > 0$, $\ln kx = \ln kx = \ln k + \ln x$.

Theorem 15. ln is invertible.

Theorem 16. For each $x > 0$, and each $n \in \mathbb{Z}^+$, $\ln x^n = n \ln x$.

Theorem 17. ln is unbounded above.

Theorem 18. For each $x > 0$, $\ln(1/x) = -\ln x$.

Theorem 19. ln is unbounded below.

Theorem 20. The image $\ln(0,\infty)$ is $\mathbb{R}$.

Definition. $\exp = \ln^{-1}: \mathbb{R} \longrightarrow \mathbb{R}$.

Theorem 21. The image $\exp \mathbb{R}$ is $(0,\infty)$.

34

8. Invertible Functions. Arc-length and Path-length.

Theorem 22. exp0 = 1.

Theorem 23. For each k, x $\in \mathbb{R}$, exp(k + x) = (expk)(expx).

Theorem 24. exp' = exp.

Definition. For each a > 0, a^x = exp(xlna).

(Note that this definition is forced on us, because we want $\ln a^x$ to be xlna.)

Theorem 25. For each a > 0, $\ln a^x$ = xlna.

For x > −1, we have 1 + x > 0. Therefore the expression $(1 + x)^{1/x}$
defines a function (−1,∞) − {0} — $\mathbb{R}$.

Theorem 26. $\lim\limits_{x \to 0} (1 + x)^{1/x} = \ln^{-1} 1$.

Definition. $e = \ln^{-1} 1$.

Theorem 27. For each x, e^x = expx.

Theorem 28. For each a > 0, a ≠ 1, let $f(x) = a^x$. Then f is a function
(−∞,∞) $\longrightarrow$ (0,∞); the image is all of (0,∞); and f is invertible.

Theorem 29. For each a > 0, a ≠ 1, $\log_a$ is the inverse of the f of
Theorem 28. That is, $y = \log_a x$ if $x = a^y$.

Theorem 30. e > 2.

Theorem 31. For each x, $\log_e x$ = lnx.

Theorem 32. $\lim\limits_{x \to \infty} \ln x = \infty$; $\lim\limits_{x \to 0^+} \ln x = -\infty$.

(The proofs need to begin with the definitions of the two limit relations.
Compare with Theorems 17 and 19 above.)

Note that we now have two different definitions of x^n, as follows.

Definition 1. For each $x \in \mathbb{R}$, $n \in \mathbb{Z}^+$,

$$x^n = f(x) = x.x...x \quad \text{(to n factors).}$$

Definition 2. For each $x \in (0,\infty)$, $n \in \mathbb{R}$,

$$x^n = g(x) = \exp(n\ln x).$$

When $x \in (0,\infty)$ and $n \in \mathbb{Z}^+$, both these definitions apply, and we need to
show that they agree. Proof?

Theorem 33. For each $x \in (0,\infty)$, $k \in \mathbb{R}$, let $f(x) = x^k$. Then $f'(x) = kx^{k-1}$.

Note that the elementary proofs of this work only when k is rational. The
first complete proof is due to Abel.

35

Problem Set 8

1. The proof of Theorem 8 would have been easier if we had been able to use the following: <u>Lemma (?)</u>. For each i there is a single point t'_i, between t_{i-1} and t_i, such that

$$P_{i-1}P_i = \sqrt{f'(t'_i)^2 + g'(t'_i)^2} \, \Delta t_i \; .$$

We could then have expressed $S(N)$ as a sample sum of a single function, and passed to the limit. But the "Lemma" is false. Give an example of a path (with f' and g' continuous) for which the lemma fails. There is a very simple example of this sort.

2. For $0 \le t \le 1$, let $f(t) = t^{17}$, $g(t) = \sqrt{1 - t^{34}}$, $p(t) = f(t), g(t))$. Find the length of p.

3. The familiar process for calculating indefinite integrals by substitution is described by the following diagram.

$$\int f(x)dx \quad \xrightarrow[\;(1)\;]{x \longrightarrow \phi(t)} \quad \int f(\phi(t))\phi'(t)dt$$

$$\| \;\; (4) \qquad\qquad\qquad\qquad\qquad\qquad\qquad \| \;\; (2)$$

$$G(\phi^{-1}(x)) + C \quad \xleftarrow[\;(3)\;]{t \longrightarrow \phi^{-1}(x)} \quad G(t) + C \; .$$

In this process, we assume that f is continuous on an interval I, that ϕ and ϕ' are continuous on an interval J, such that $\phi(J) = I$. What else do we need to assume, to show that $G(\phi^{-1}(x)) + C$ is the answer to the integration problem that we started with? Note that the use of differentials in the above diagram is merely conventional and convenient, with $dx \longrightarrow \phi'(t)dt$ when $x \longrightarrow \phi(t)$. The validity of the process depends not on a statement about differentials, but on a statement about derivatives, namely:

$$G' = f(\phi)\phi' \implies [G(\phi^{-1})]' = f.$$

4. Suppose that in Theorem 4 we had used the additional hypothesis that f is a function $\mathbb{R} \twoheadrightarrow \mathbb{R}$. Questions:

(a) Would this have made the proof any easier?

(b) Is the answer to (a) of any significance?

Chapter 9: Point-wise Convergence and Uniform Convergence

<u>Definition</u>. Let f_1, f_2, ... and f be functions $A \to \mathbb{R}$, where $A \subset \mathbb{R}$. Suppose that for each $x \in A$ we have

$$\lim_{n \to \infty} f_n(x) = f(x).$$

Then the sequence f_1, f_2, ... <u>converges point-wise to</u> f.

In the following descriptions of examples, it should be understood that $A = [a,b]$, and that f_1, f_2, ... converges point-wise to f.

<u>Example 1</u>. Each f_n is continuous, but f is not continuous.

<u>Example 2</u>. Each f_n is integrable, and $|f_n(x)| \leq 1$ for each n and each x, but f is not integrable.

<u>Example 3</u>. Each f_n is integrable, and f is integrable, but $\int_a^b f_n$ does not approach a limit, as $n \to \infty$.

<u>Example 4</u>. Each f_n is continuous, f is continuous, and $\lim_{n \to \infty} \int_a^b f_n$ exists, but $\lim_{n \to \infty} \int_a^b f_n \neq \int_a^b f$.

The last straw is the following.

<u>Example 5</u>. $f(x) = 0$ for each x, but every interval in $\mathbb{R}$ contains numbers of the form $\int_a^b f_n$. That is, every real number is a limit-point of the set $\{\int_a^b f_n\}$.

Evidently these examples are a description of a morass: point-wise convergence is not related either to continuity or to integration in any reasonable way. Two ways have been found, to get around these troubles.

(I). We can define the integral differently, in such a way that it does not depend on continuity, or on continuity almost everywhere. Such a new definition was given by Henri Lebesgue. It is too complicated to be worth stating in a course -- such as this one -- in which it will not be used. It is easy, however, to state some of the important properties of the Lebesgue integral.

1. To be Lebesgue-integrable, a function need not be continuous anywhere at all. For example, consider $f: [0,1] \to \mathbb{R}$, where $f(x) = 0$ when x is rational and $f(x) = 1$ when x is irrational. Then f is Lebesgue-integrable, and its Lebesgue integral is

$$(L) \int_a^b f = 1.$$

Thus, in effect, the Lebesgue integral ignores null sets, as if they were empty; the rational numbers on [0,1] form a null set, and we are integrating f as if we had f(x) = 1 for each x.

2. Let f_1, f_2, ... be a sequence of non-negative Lebesgue-integrable functions [a,b] → ℝ, with a common bound M, so that $0 \le f(x) \le M$ for each x. If the sequence converges point-wise to a function f, then f is Lebesgue-integrable, and the integral of the limit is the limit of the integrals. (In fact, a more general result holds, but its statement is complicated.)

3. If f: [a,b] → ℝ is integrable in the sense of Riemann (as defined in Section 6), then f is Lebesgue-integrable, and

$$\int_a^b f = (L) \int_a^b f .$$

(II). The Lebesgue theory is elegant and powerful, and for many purposes it is needed. But for many other purposes a much simpler remedy is adequate and more appropriate: we can retain the Riemann integral, but give a more restrictive definition of convergence, for sequences of functions, as follows.

Definition. Let f_1, f_2, ... and f be functions A → ℝ, where A ⊂ ℝ. Suppose that for every ε > 0 there is an $n_\varepsilon \in \mathbb{Z}^+$ such that

$$x \in A \quad \text{and} \quad n \ge n_\varepsilon \implies |f_n(x) - f(x)| < \varepsilon .$$

Then we say that the sequence f_1, f_2, ... converges uniformly to f, and we write

$$\text{ULim}_{n \to \infty} f_n = f \qquad \text{(on A)}.$$

Here the use of the term uniformly is natural; the point is that n_ε depends only on ε, and not on x. Compare with the definition of uniform continuity, for functions.

Theorem 1. Suppose that $\text{ULim}_{n \to \infty} f_n = f$ on [a,b], let $x_0 \in$ [a,b], and suppose that each f_n is continuous at x_0. Then f is continuous at x_0.

We get an obvious corollary.

Theorem 2. If f_1, f_2, ... are all continuous, and $\text{ULim}_{n \to \infty} f_n = f$ on [a,b], then f is continuous and integrable on [a,b].

But to discuss $\int_a^b \text{ULim}_{n \to \infty} f_n$, we need to know a little more about integrals, as follows.

Theorem 3. If f is integrable on [a,b], and k ∈ ℝ, then kf is integrable on [a,b], and $\int_a^b kf = k \int_a^b f$.

9. Point-wise Convergence and Uniform Convergence.

Theorem 4. If f and g are integrable on $[a,b]$, then $f + g$ is integrable on $[a,b]$, and

$$\int_a^b (f + g) = \int_a^b f + \int_a^b g.$$

Theorem 5. If f is integrable on $[a,b]$, then $|f|$ is integrable on $[a,b]$, and

$$\left| \int_a^b f \right| \leq \int_a^b |f| .$$

Theorem 6. If $\underset{n \to \infty}{\text{ULim}} f_n = f$ on $[a,b]$, and for each n, f_n is continuous, then

$$\int_a^b f = \underset{n \to \infty}{\text{Lim}} \int_a^b f_n .$$

Theorems 2 and 6 are adequate for the purposes of this course. At the price of more technique, they can be extended in the following ways.

Let f_1, f_2, ... be a sequence of functions $[a,b] \to \mathbb{R}$. Suppose that there is an $M > 0$ such that $|f_n(x)| \leq M$ for each n and each x. Then the sequence is <u>uniformly</u> <u>bounded</u>.

Theorem 7. Let f_1, f_2, ... and f be functions $[a,b] \to \mathbb{R}$, such that

$$\underset{n \to \infty}{\text{ULim}} f_n = f \qquad (\text{on } [a,b]).$$

If each f_n is bounded, then the sequence f_1, f_2, ... is uniformly bounded, and f is bounded.

Theorem 8. Let f_1, f_2, ... and f be as in Theorem 7. If each f_n is integrable, then f is integrable.

Theorem 9. Under the conditions of Theorem 7 and 8, $\underset{n \to \infty}{\text{Lim}} \int_a^b f_n = \int_a^b f.$

Chapter 10: Infinite Series

Let a_1, a_2, ... be a sequence of real numbers. For each n, let

$$A_n = \sum_{i=1}^{n} a_i \; .$$

If $\underset{n \to \infty}{\text{Lim}} A_n$ exists, $= A$, then we say that the series $\sum_{i=1}^{\infty} a_i$ is __convergent__, and we write

$$\sum_{i=1}^{\infty} a_i = A \; .$$

We shall make minor variations on this definition without comment. For example, we may discuss $\sum_{i=0}^{\infty} a_i$, or $\sum_{i=k}^{\infty} a_i$, and so on. A series which does not converge is called __divergent__.

Let A_1, A_2, ... be a sequence. Suppose that for every $M > 0$ there is an $n_0 \in \mathbb{Z}^{+}$ such that $n \geq n_0 \implies A_n > M$. Then we say that A_1, A_2, ... __diverges to infinity__, and we write

$$\underset{n \to \infty}{\text{Lim}} A_n = \infty \; .$$

Similarly for $\underset{n \to \infty}{\text{Lim}} A_n = -\infty$. Note that if $A_n \longrightarrow \infty$, the sequence is __not__ called convergent. If $A_n = \sum_{i=1}^{n} a_i$, as before, and $\underset{n \to \infty}{\text{Lim}} A_n = \infty$ (or $-\infty$), then we write

$$\sum a_i = \infty \qquad (\text{or} \; \sum a_i = -\infty) \; ,$$

and we say that $\sum a_i$ __diverges to infinity__ (or to $-\infty$).

At this point, it would be worthwhile to re-examine Sections 4 and 5, recalling various theorems on sequences. In particular, Theorem 4.12 asserts that every convergent sequence is bounded; and Theorem 5.7 asserts that if $0 \leq r < 1$, then $\underset{n \to \infty}{\text{Lim}} r^n = 0$.

__Theorem 1.__   If $|r| < 1$, then $\underset{n \to \infty}{\text{Lim}} r^n = 0$.

__Theorem 2.__   If $r \neq 1$, then

$$\sum_{i=0}^{n} r^i = (r^{n+1}-1)/(r-1) \; .$$

__Theorem 3.__   If $|r| < 1$, and $a \in \mathbb{R}$, then

$$\sum_{i=0}^{\infty} ar^i = a/(1-r) \; .$$

10. Infinite Series

Here and hereafter, in interpreting summations, we agree that $r^0 = 1$, even when $r = 0$. This saves us the bother of writing such expressions as $a + \sum\limits_{i=1}^{\infty} ar^i$. The convention $0^0 = 1$ is a convention of notation; it is not intended to have any mathematical meaning.

<u>Theorem 4</u>. If $\sum a_i$ is convergent, then $\lim\limits_{i \to \infty} a_i = 0$.

<u>Example 1</u>. Given that $\lim\limits_{i \to \infty} a_i = 0$, it does not follow that $\sum a_i$ is convergent.

There is a very rudimentary example of this sort. We do not need anything so subtle as:

<u>Theorem 5</u>. $\sum\limits_{i=1}^{\infty} (1/i) = \infty$.

If $a_i \geq 0$ for each i, then either (1) $\sum a_i$ converges, to a finite sum or (2) $\sum a_i = \infty$. Therefore we may indicate (1) briefly by writing

$$\sum a_i < \infty \qquad\qquad (a_i \geq 0).$$

<u>Theorem 6</u>.

$$\sum\limits_{i=1}^{\infty} \frac{1}{i^2} < \infty .$$

<u>Theorem 7</u>. (The Comparison Theorem for Positive Series.) Given $\sum a_i$, $\sum b_i$, where $a_i \geq 0$, $b_i \geq 0$. Suppose that there is an n_0 such that

$$n \geq n_0 \implies a_i \leq b_i .$$

Then (1) $\sum b_i < \infty \implies \sum a_i < \infty$ and (2) $\sum a_i = \infty \implies \sum b_i = \infty$.

As usual, for $n \in \mathbb{Z}^+$ we define

$$n! = 1 \cdot 2 \cdot 3 \ldots n .$$

For convenience in writing summations, we use the convention $0! = 1$.

<u>Theorem 8</u>.

$$\sum\limits_{n=0}^{\infty} \frac{1}{n!} < \infty .$$

<u>Theorem 9</u>. (The Weierstrass M-test for Uniform Convergence.) Let A be a non-empty set of real numbers, and let f_1, f_2, ... and f be functions $A \to \mathbb{R}$. Then (1) $\text{ULim}\limits_{n \to \infty} f_n = f$ on A if and only if (2) there is a sequence M_1, M_2, ... of positive constants such that (a) $\lim\limits_{n \to \infty} M_n = 0$ and (b) for each n, and each $x \in A$, $|f_n(x) - f(x)| \leq M_n$.

Problem Set 10

1. We rewrite Theorem 2 in the form:

$$\sum_{i=0}^{\infty} x^i = 1/(1-x) \qquad (|x| < 1).$$

For each n, let $A_n(x) = \sum_{i=0}^{n} x^i$. Show that for $0 < k < 1$,

$$\underset{n \to \infty}{\text{ULim }} A_n(x) = 1/(1-x) \qquad \text{on } [-k,k].$$

Then show that the convergence is not uniform on the open interval $(-1,1)$.

2. For each n, let $B_n(x) = \sum_{i=0}^{n} (-1)^i x^i$. Show that (a) for $0 < k < 1$,

$$\underset{n \to \infty}{\text{ULim }} B_n(x) = 1/(1+x) \qquad \text{on } [-k,k].$$

Show that (b) the convergence is not uniform on $(-1,1)$. Then show that (c) for $x < 1$,

$$\ln(1+x) = \sum_{i=1}^{\infty} (-1)^{i+1} x^i/i.$$

(Note that if we proved, somehow, that the series on the right converges, this would not constitute of proof of (c). Note, however, that (c) automatically means that the series on the right converges.)

3. For each n, let $C_n(x) = \sum_{i=0}^{n} (-1)^i x^{2i}$. Show that if $0 < k < 1$, then

$$\underset{n \to \infty}{\text{ULim }} C_n(x) = 1/(1+x^2) \qquad \text{on } [-k,k].$$

Then show that for $|x| < 1$,

$$\text{Tan}^{-1} x = \sum_{i=0}^{\infty} (-1)^i x^{2i+1}/(2i+1).$$

Here Tan is the restriction of the function $f(x) = \tan x$ to the open interval $(-\pi/2, \pi/2)$, and Tan^{-1} is (as usual) the inverse of Tan. The comment on Problem 2 also applies to Problem 3.

4. Suppose that $\sum_{i=0}^{\infty} a_i x^i$ converges for every x. Then the series defines a function f. It will turn out that functions expressible in this way are always differentiable, with

$$f'(x) = \sum_{i=1}^{\infty} i a_i x^{i-1} .$$

Granted that this is true, what must the numbers a_i be, if (1) $f(0) = 1$ and (2) $f'(x) = f(x)$ for each x?

5. Show that $\exp (\exp x = e^x$ for each x) is the only function which is defined for each x and satisfies conditions (1) and (2) of Problem 4.

6. A naive solution of Problem 5 might run as follows:

$$dy/dx = y; \quad dy/y = dx; \quad \int dy/y = \int dx; \quad \ln y = x + C;$$
$$y = e^{x+C}; \quad y = 1 \quad \text{when} \quad x = 0; \quad C = 0; \quad y = e^x.$$

This "proof" raises various questions, including, at least, the following. Evidently the symbol "dy/y" is the name of some sort of mathematical object. What sort? Is dy/y a function? If so, of how many variables? Is dy/y a variable? In either case, what exactly do we mean by the term <u>variable</u>? You may find it harder to answer questions like this than to prove the simple theorem stated in Problem 5.

Moreover, the "proof" above is defective in a simple way which has nothing to do with any complex question. What is the simple defect?

7. Consider the formula

$$\int x^n dx = x^{n+1}/(n+1) + C \qquad (n \neq -1).$$

This appears to mean that if $n \neq -1$, and $f'(x) = x^n$, then $f(x)$ is of the form $x^{n+1}/(n+1) + C$, where C is a constant. Discuss.

8. Is the following true? <u>Theorem (?)</u>. For $0 < s < 1$, $\sum\limits_{i=1}^{\infty} is^i < \infty$.

9. Let $k \in \mathbb{R}$. Then for each $n \in \mathbb{Z}^+$,

$$\sum_{i=1}^{n} \frac{1}{i^k} \leq 1 + \int_1^n \frac{dx}{x^k} \quad,$$

and

$$\sum_{i=1}^{n} \frac{1}{i^k} \leq \int_1^{n+1} \frac{dx}{x^k} \quad.$$

Chapter 11: Absolute Convergence. Rearrangements of Series

Definition. If the series $\sum |a_i|$ is convergent, then the series $\sum a_i$ is absolutely convergent.

Theorem 1. If $\sum |a_i| < \infty$, then $\sum a_i$ is convergent.

Theorem 2. If $\sum |a_i| < \infty$, then $|\sum a_i| \leq \sum |a_i|$.

Thus absolute convergence implies convergence. But the converse is false; and in fact we have a sort of "counter-theorem", as follows.

Theorem 4. (The Alternating Series Test.) Let a_1, a_2, $\ldots$ be a sequence of positive numbers, with $a_{i+1} < a_i$ for each i, and $\operatorname*{Lim}_{i \to \infty} a_i = 0$. Then

$$\sum_{i=1}^{\infty} (-1)^{i+1} a_i \quad \text{converges.}$$

Theorem 5. Under the conditions of Theorem 4, let

$$A_n = \sum_{i=1}^{n} (-1)^{i+1} a_i, \; A = \sum_{i=1}^{\infty} (-1)^{i+1} a_i .$$

Then for each k we have

$$|A_n - A| < a_{n+1} .$$

Thus, if we use A_n as an approximation of A, then the error is less than the absolute value of the first term that we do not use.

Theorem 1 has a variety of applications, because except for Theorem 4 most of our convergence tests apply only to series with positive terms; this is true of comparison tests (Theorem 10.7) and also of integral tests. (See Problem 9.19). Theorem 6. (The Ratio Test). Given $\sum_{i=1}^{\infty} a_i$, where $a_i \neq 0$ for each i. Let $r_n = |a_{n+1}/a_n|$. Then $\sum a_i$ is convergent.

Note that this proves absolute convergence whenever it applies at all. For series in general, it may seem almost a miracle for r_n to approach any limit whatever; but in fact this happens very often for series defined by simple formulas and recursion processes. For example, the RationTest trivializes Theorem 10.8:

$$\sum 1/n! < \infty.$$

11. Absolute Convergence. Rearrangements of Series.

   Definition. Let n_1, n_2, ... be a sequence of positive integers, such that each positive integer k is $= n_i$ for one and only one i. (That is, the function $i \longmapsto n_i$ is a bijection $\mathbb{Z}^+ \to \mathbb{Z}^+$.) Let $\sum_{i=1}^{\infty} a_i$ be a series. Then $\sum_{i=1}^{\infty} a_{n_i}$ is a rearrangement of $\sum_{i=1}^{\infty} a_i$.

   Theorem 7. If $\sum a_i$ is absolutely convergent, then every rearrangement of $\sum a_i$ is convergent, and has the same sum.

   Thus we have a "law of total commutativity" for absolutely convergent series. But there is no such law for convergent series in general; we have counter-theorems as follows.

   Theorem 8. Let $\sum a_i$ be a series which is convergent but not absolutely convergent, let $\sum a_{n_i}$ be a rearrangement of $\sum a_i$ and for each n, let

$$A_n' = \sum_{i=1}^{n} a_{n_i} . \quad \text{Then}$$

   (a) For each $k \in \mathbb{R}$, the rearrangement can be chosen so that $\sum a_{n_i} = k$.

   (b) The rearrangement can be chosen so that $\sum a_{n_i} = \infty$.

   (c) $\sum a_{n_i}$ can be chosen so that $\sum a_{n_i} = -\infty$.

   (d) For each interval I in $\mathbb{R}$, $\sum a_{n_i}$ can be chosen so that every sub-interval J of I contains some number A_n' .

Problem Set 11

1. Find a rearrangement of $\sum_{i=1}^{\infty} (-1)^{i+1}(1/i)$ which converges to 0.

2. Investigate for convergence and for absolute convergence.

$$\sum_{i=0}^{\infty} \frac{\sin(3i^3 + i^2+1)}{i!} .$$

3. If $a_i \geq 0$ for each i, then $\sum a_i < \infty$ if and only if the partial sums

$$A_n = \sum_{i=1}^{n} a_i \quad \text{form a bounded set.}$$

Chapter 12: Power Series

A <u>power</u> series is a series of the form $\sum_{i=0}^{\infty} a_i x^i$, where $a_i \in \mathbb{R}$ for each i.
Evidently the convergence of a power series may depend on the value assigned to x.
Evidently every such series converges for $x = 0$.

    <u>Example 1</u>. There is a power series which converges for every x.

    <u>Example 2</u>. There is a power series which diverges for every $x \neq 0$.

    We shall now see what happens when a power series converges for some $x_0 \neq 0$,
but not necessarily for every x.

    <u>Theorem 1</u>. If $\sum_{i=0}^{\infty} a_i x_0^i$ is convergent, and $|x| < |x_0|$, then $\sum_{i=0}^{\infty}$ is ab-
solutely convergent. In fact, if $M \in \mathbb{R}$ and $|a_i x_0^i| \leq M$ for each i, then

$$\sum_{i=0}^{\infty} |a_i x^i| \leq M \; \frac{|x_0|}{|x_0| - |x|} \qquad (|x| < |x_0|).$$

    <u>Theorem 2</u>. If $\sum_{i=0}^{\infty} a_i x^i$ converges for some $x_0 \neq 0$, but diverges for some
$x_2 \neq 0$, then the series converges for some $x_3 > 0$, but not for every $x > 0$.

    <u>Theorem 3</u>. Under the conditions of Theorem 2, there is a number R such that
(1) the series converges for $|x| < R$ and (2) the series diverges for $|x| > R$.

    The number R is evidently unique. It is called the <u>radius of convergence</u>
of the series.

    Note that in the light of Theorem 3, there is no power series which converges
on the integral $(-1,2)$ and diverges on the intervals $(-\infty, -1)$ and $(2,\infty)$. But
the behavior of the series at R and $-R$ need not be symmetric.

    <u>Examples</u>. Let C be the set on which the series $\sum a_i x^i$ converges.
Then C may be of any of the following types: $(-\infty,\infty)$, $\{0\}$, $(-R,R)$, $[-R,R)$, $(-R,R]$,
and $[-R,R]$.

    <u>Theorem 4</u>. Suppose that $\sum_{i=0}^{\infty} a_i x^i$ converges on $(-R,R)$, and let k be a number
between 0 and R. For each n, let $A_n(x) = \sum_{i=0}^{n} a_i x^i$.

Then

$$\underset{n \to \infty}{\mathrm{ULim}} \; A_n(x) = \sum_{n=1}^{\infty} a_i x^i \qquad \text{on } [-k,k] .$$

    <u>Theorem 5</u>. Let $f(x) = \sum_{i=0}^{\infty} a_i x^i$, for $x \in (-R,R)$. Then f is continuous.

Theorem 6.   Under the conditions of Theorem 4, for $-R < a < b < R$,

$$\int_a^b f(x) \; dx = \int_a^b \sum_{i=0}^{\infty} a_i x^i = \sum_{i=0}^{\infty} \int_a^b a_i x^i \; dx \; .$$

Theorem 7. Under the conditions of Theorems 5 and 6, for each $x \in (-R,R)$ we have

$$\int_0^x f(t) \; dt = \sum_{i=0}^{\infty} \int_0^x a_i t^i \; dt = \sum_{i=0}^{\infty} \frac{a_i x^{i+1}}{i+1} \; .$$

(Compare with Problems 10.2 and 10.3.)

Theorem 8.   If $\sum_{i=0}^{\infty} a_i x^i$ converges on $(-R,R)$, then so also does $\sum_{i=1}^{\infty} i \; a_i x^{i-1}$.

Theorem 9.   If $\sum_{i=0}^{\infty} a_i x^i$ converges to a function f, on $(-R,R)$, then f is differentiable on $(-R,R)$, and

$$f'(x) = \sum_{i=1}^{\infty} i a_i x^{i-1} \qquad (x \in (-R,R)) \; .$$

If the function f has an ith derivative, then the ith derivative is denoted by $f^{(i)}$.

Theorem 10.   If $f(x) = \sum_{i=0}^{\infty} a_i x^i$ on $(-R,R)$, then f has derivatives of all orders on $(-R,R)$; and for each i,

$$f^{(i)}(0) = i! a_i \; ,$$

so that for each i,

$$a_i = \frac{f^{(i)}(0)}{i!} \; .$$

Theorem 11.   If $f(x) = \sum_{i=0}^{\infty} a_i x^i = \sum_{i=0}^{\infty} b_i x^i$ on $(-R,R)$, then for each i, $a_i = b_i$.

Thus power series behave like (finite) polynomials: there are no non-trivial identities between them, and so two of them give the same function only when they obviously must.

Note that in Theorems 4 - 11 we have required that our series converge on an interval $(-R,R)$, but we have not required that the series diverge for $|x| > R$. Therefore Theorems 4 - 11 apply, as stated, to a series which converges for every x.

Problem Set 12

1. In Theorem 9, the series $\sum_{i=1}^{\infty} ia_i x^{i-1}$ is not in "standard form": the exponent of x is not the same as the index of summation. Express this in "standard form", as $\sum_{j=?}^{\infty} b_j x^j$. We will then have $f'(x) = \sum_{i=?}^{\infty} b_i x^i$.

2. Iterating the process of Theorem 10, we get

$$f''(x) = \sum_{i=2}^{\infty} i(i-1)x^{i-2} .$$

Express this in a "standard form" $\sum_{i=?}^{\infty} c_i x^i$.

3. The converse of Theorem 10 is false: if a function f is defined for every x, and has derivatives of all orders, it does not follow that f is the sum of a power series. The classical example of this is

$$f(x) = \begin{cases} e^{-1/x^2} & \text{for } x \neq 0 \\ \\ 0 & \text{for } x = 0. \end{cases}$$

It is evident that every $x \neq 0$, f has derivatives of all orders. Show that for each i, $f^{(i)}(0)$ is well-defined, and is $= 0$. Suppose now that f has a power series expansion, with

$$f(x) = \sum_{i=0}^{\infty} a_i x^i .$$

Then $a_i = 0$ for each i, and so $f(x) = 0$ for each x, which is false.

Chapter 13: Power Series for Elementary Functions

Theorem 1. Suppose that f is a function $\mathbb{R} \longrightarrow \mathbb{R}$, such that (1) f is differentiable, with $f' = f$, (2) $f(0) = 1$, and (3) there is a series $\sum\limits_{i=0}^{\infty} a_i x^i$ which converges to f. Then for each i we have $a_i = 1/i!$.

Theorem 2. For each x, $\sum\limits_{i=0}^{\infty} \dfrac{x^i}{i!}$ converges, to a function g; $g'(x) = g(x)$ for each x; and $g(0) = 1$.

Theorem 3. For each x, $e^x = \sum\limits_{i=0}^{\infty} \dfrac{x^i}{i!}$.

Theorem 4. Let f be a function $(-1,1) \rightarrow \mathbb{R}$, such that (1) $f(0) = 1$, (2) $f(x)$ is the sum of a series $\sum\limits_{i=0}^{\infty} a_i x^i$ ($|x| < 1$), and (3) for each $x \in (-1,1)$, $(x+1)f'(x) = k\, f(x)$ ($k \in \mathbb{R}$). Then for each i,

$$a_i = \binom{k}{i} = \frac{k(k-1)\ldots(k-i+1)}{i!} \quad .$$

Theorem 5. For each $x \in (-1,1)$,

$$\sum\limits_{i=0}^{\infty} \binom{k}{i} x^i$$

converges, to a function g; $g(0) = 1$, and for each $x \in (-1,1)$, $(x+1)g'(x) = kg(x)$.

Theorem 6. For each $x \in (-1,1)$, and each $k \in \mathbb{R}$,

$$(1 + x)^k = \sum\limits_{i=0}^{\infty} (-1)^{i+1} \frac{x^{2i+1}}{(2i+1)!} \quad .$$

Theorem 7. Let f be a differentiable function $\mathbb{R} \rightarrow \mathbb{R}$, such that (1) $f(0) = 0$, (2) $f'(0) = 1$, (3) $f'' = -f$, and (4) $f(x)$ is the sum of a series of the form $\sum\limits_{i=0}^{\infty} a_i x^i$. Then for each i, $a_{2i} = 0$, and

$$f(x) = \sum\limits_{i=0}^{\infty} (-1)^{i+1} \frac{x^{2i+1}}{(2i+1)!} \quad .$$

Analysis

Theorem 8. For each x, the series

$$\sum_{i=0}^{\infty} (-1)^{i+1} \frac{x^{2i+1}}{(2i+1)!}$$

converges, to a function g; and g satisfies the conditions for f in Theorem 7.

Theorem 9. Let g be as in Theorem 8; and let h = g'. Then for each x,

$$h(x) = \sum_{i=0}^{\infty} (-1)^i \frac{x^{2i}}{(2i)!} \; ;$$

h(0) = 1, and h'(0) = 0.

Theorem 10. Let g and h be as in Theorems 8 and 9. Then for each x, g(x) = sin x and h(x) = cos x.

(In proving Theorems 3 and 6, you probably showed that a certain function was constant and = 1. To prove Theorem 10, you may find it a good idea to show that a certain function is constant and = 0.)

Theorem 11. For each n $\in \mathbb{Z}^+$, and each a, b $\in \mathbb{R}$,

$$(a + b)^n = \sum_{i=0}^{n} \binom{n}{i} a^i b^{n-i} \quad .$$

(This is, of course, the elementary binomial theorem. For two reasons, to deduce it from Theorem 6 deserves to be regarded as a prank. First, it has an elementary proof, by induction. Second, the elementary proof is much more general: it shows that the formula holds not just in $\mathbb{R}$ but in any ring. A ring is a system $[R, +, \cdot]$ which satisfies all of the field postulates, except perhaps for the postulate which states that every number different from 0 has a reciprocal.)

The methods called for in this section are examples of methods which apply much more generally, in investigating differential equations for which no solution is known at the outset.

Topology

Chapter 1: Sets and Functions

We shall use the standard terms and notations of analysis and set theory. (Thus much of the following has already appeared in the first few pages of <u>Analysis</u>.) $\mathbb{R}$ is the set of all real numbers, and $\mathbb{Z}$ is the set of all integers. If A is a set, then $x \in A$ means that x belongs to A. If x does not belong to A, then we write $x \notin A$. Let S be a set. Then

$$\{x | x \in S \text{ and } (\dots)\}$$

denotes the set of all elements of S that satisfy the condition $(\dots)$. Thus the <u>union</u> of two sets A and B is

$$A \cup B = \{x | x \in A \text{ or } x \in B\}.$$

The <u>intersection</u> of A and B is the set

$$A \cap B = \{x | x \in A \text{ and } x \in B\}.$$

If $x \in A$ and $x \in B$ never both hold, then A and B are <u>disjoint</u>, and we write

$$A \cap B = \emptyset ,$$

where $\emptyset$ is the empty set. If P and Q are propositions, then

$$P \Rightarrow Q$$

means that P implies Q; that is, if P, then Q. If $P \Rightarrow Q$ and also $Q \Rightarrow P$, then we say that P and Q are <u>equivalent</u>, and we write

$$P \Leftrightarrow Q .$$

Let A and B be sets. If for every x, $x \in A \Rightarrow x \in B$, then A is a <u>subset</u> of B, and we write $A \subset B$. (Here $A \subset A$ for every set A. Thus $\subset$ is like $\leq$, not like $<$.) If $A \subset B$, then we may also write $B \supset A$. The <u>difference</u> of the sets A and B is the set

$$A - B = \{x | x \in A \text{ and } x \notin B\}.$$

Here it is not assumed that $A \subset B$. Thus $A - B$ is well defined for all sets A and B. The symbol $\{a\}$ denotes the singleton whose only element is a; and $\{a, b, c, \dots\}$ denotes the set whose elements are a, b, c, $\dots$. By abuse of language, we may write $A - x$ to mean $A - \{x\}$.

The <u>ordered pair</u> with a as first term and b as second term is denoted by (a,b). If A and B are sets, then the <u>product</u> $A \times B$ is the set of all ordered pairs of the form (a,b), with $a \in A$ and $b \in B$. We may write A^2 for $A \times A$.

More generally, for each positive integer   n,

$$A^n = A \times A \times \ldots \times A \qquad \text{(to   n   factors.)}$$

$\mathbb{Z}^+$ is the set of all positive integers.

A <u>function</u>

$$f: A \longrightarrow B$$

of   A   into   B   is a triplet   [A,B,f], where   A   and   B   are non–empty sets and   f   is a collection of ordered pairs such that (1) if $(a,b) \in f$, then $a \in A$ and $b \in B$ and (2) each element   a   of   A   is the first term of one and only one element   (a,b)   of   f.   A   is the <u>domain</u> of the function, and   B   is the <u>codomain</u>.   If $(a,b) \in f$, then we write

$$b = f(a), \text{ or } a \longmapsto b.$$

The <u>image</u> is the set

$$f(A) = \{b \mid b \in B \text{ and } b = f(a) \text{ for some } a \in A\}.$$

More generally, for each $A' \subset A$, <u>the image of</u>   A'   is the set

$$f(A') = \{b \mid b \in B \text{ and } b = f(a) \text{ for some } a \in A'\}.$$

If $a \neq a' \Rightarrow f(a) \neq f(a')$, then   f   is <u>injective</u>, and is an <u>injection</u>.   If   f(A) is all of   B, then   f   is <u>surjective</u>, and is a <u>surjection</u>.   If   f   is both injective and surjective, then   f   is <u>bijective</u>, and is a <u>bijection</u>.

Given $f: A \longrightarrow B$, by abuse of language we may refer to   f   as a function. Strictly speaking, however, the codomain   B   is part of the definition of the function; if not, the term <u>surjective</u> would have no meaning.   (Any function becomes surjective, if we redefine the codomain so as to make it the image.)

If $f: A \longrightarrow B$ is a bijection, then we write

$$f: A \longleftrightarrow B .$$

In this case, there is a bijection f^{-1}, called the <u>inverse</u> of   f, such that $f^{-1}(f(a)) = a$ for each $a \in A$ and $f(f^{-1}(b)) = b$ for each $b \in B$.   To be precise,

$$f^{-1} = \{(b,a) \mid (b,a) \in B \times A \text{ and } (a,b) \in f\} .$$

Whether or not   f   has an inverse, we define the <u>inverse image</u> of each $b \in B$ to be

$$f^{-1}(b) = \{a \mid a \in A \text{ and } f(a) = b\}.$$

Note that $f^{-1}(b)$ is a subset of   A, and not an element of   A.   And $f^{-1}(b) = \emptyset$ whenever $b \notin f(A)$.   More generally, for each $B' \subset B$,

$$f^{-1}(B') = \{a \mid a \in A \text{ and } f(a) \in B'\} .$$

Let $f: A \longrightarrow B$ be a function, and let   A'   be a subset of   A.   Consider the function $g: A' \longrightarrow B$ defined by the condition   g(a) = f(a)   for every $a \in A'$.   The function g is called the <u>restriction of</u> f <u>to the set</u> A', and is denoted by $f|A'$.

Let A be a set. A <u>sequence of elements of</u> A is a function

$$f: \mathbb{Z}^+ \longrightarrow A \ .$$

Ordinarily, we use the notation a_i for $f(i)$, and write the sequence in the form

$$a_1, \ a_2, \ \ldots \ .$$

Note that the sequence is a different object from the image set

$$\{a_1, \ a_2, \ \ldots\} = \{a \mid a = a_i \ \text{for some} \ i \in \mathbb{Z}^+\} \ .$$

If M_1, M_2, $\ldots$ is a sequence of sets, then

$$\bigcap_{i=1}^{\infty} M_i$$

is the intersection of all the sets M_i. That is,

$$\bigcap_{i=1}^{\infty} M_i = \{P \mid P \in M_i \ \text{for each} \ i\} \ .$$

Similarly,

$$\bigcup_{i=1}^{\infty} M_i$$

is the union of all the sets M_i. We use similar notations

$$\bigcap_{i=1}^{n} M_i \ , \ \bigcup_{i=1}^{n} M_i$$

for the intersection and union of a finite sequence of sets. More generally, let A and G be sets, and let

$$f: A \longrightarrow G$$

$$: \alpha \longmapsto g_\alpha \in G$$

be a function. We then call A an <u>indexing set</u> for G, and we write

$$G = \{g_\alpha\} = \{g_\alpha\}_{\alpha \in A} \ .$$

We then define

$$\bigcap_{\alpha \in A} g_\alpha = \{P \mid P \in g_\alpha \ \text{for each} \ \alpha \in A\},$$

$$\bigcup_{\alpha \in A} g_\alpha = \{P \mid P \in g_\alpha \ \text{for some} \ \alpha \in A\} \ .$$

Sequences and finite sequences are included in this scheme, as the cases $A = \mathbb{Z}^+$ and $A = I_n$. Any set G is indexed by itself, under the identity function $g \longmapsto g$.

Thus the intersection and union of all elements of G can be written as

$$\bigcap_{g \in G} g \, , \quad \bigcup_{g \in G} g \, .$$

Sometimes the union of the elements of G may be denoted by G*.

The symbol $\ni$ stands for the phrase <u>such</u> <u>that</u>. It will not be used in the text, but is convenient on blackboards and in notebooks.

Chapter 2: Metric Spaces

We recall that $\mathbb{R}$ is the set of all real numbers. For each x, y in $\mathbb{R}$, let $d(x,y) = |x - y|$. Thus we have a function $d: \mathbb{R} \times \mathbb{R} \to \mathbb{R}$.

Theorem 1. The function $d: \mathbb{R} \times \mathbb{R} \to \mathbb{R}$ has the following properties:

D.1. $d(x,y) \geq 0$, for every x and y.

D.2. $d(x,y) = 0$ if and only if $x = y$.

D.3. $d(x,y) = d(y,x)$, for every x and y.

D.4. (The Triangular Inequality.) For every x, y, and z,

$$d(x,y) + d(y,z) \geq d(x,z).$$

Let E be a Euclidean plane. We choose a unit of distance, once for all, so that the distance between any two points P and Q is a well defined real number $d(P,Q)$.

Theorem 2. The function $d: E \times E \to \mathbb{R}$ has the following properties:

D.1. $d(P,Q) \geq 0$, always.

D.2. $d(P,Q) = 0$ if and only if $P = Q$.

D.3. $d(P,Q) = d(Q,P)$, always.

D.4. $d(P,Q) + d(Q,R) \geq d(P,R)$, always.

More generally, let X be any non-empty set, and let d be a function $X \times X \to \mathbb{R}$, satisfying conditions D.1–D.4. Then d is called a <u>distance function</u> for X, and the pair $[X,d]$ is called a <u>metric space</u>. Sometimes, by abuse of language, we may refer to X as a metric space, if it is clear what distance function is intended. But if d and d' are different distance functions, then $[X,d]$ and $[X,d']$ are different metric spaces.

For each positive integer n, $\mathbb{R}^n$ is the set of all ordered n-tuples $(x_1, x_2, \ldots, x_n)$ of real numbers. The <u>distance</u> between the points

$$P = (x_1, x_2, \ldots x_n), \quad Q = (y_1, y_2, \ldots y_n)$$

is defined by the formula

$$d(P,Q) = \sqrt{\sum_{i=1}^{n} (x_i - y_i)^2}.$$

Theorem 3. $[R^n, d]$ is a metric space.

For each n, the distance function for $\mathbb{R}^n$ defined by the above formula is called the <u>Cartesian</u> distance function, and the metric space $[R^n, d]$ is called <u>Cartesian n-space</u>. (In this course, we shall not be dealing with $\mathbb{R}^n$ as a vector space, or as an inner product space.)

Problem Set 2

Prove or disprove:

1. Let [X,d] be a metric space. For each P,Q in X, let d'(P,Q) = $[d(P,Q)]^2$. (Briefly, d' = d^2.) Then [X,d'] is a metric space.

2. Let [X,d] be a metric space. For each P,Q in X, let d'(P,Q) = $\sqrt{d(P,Q)}$. (Briefly, d' = $\sqrt{d}$.) Then [X,d'] is a metric space.

3. Let S^2 be the unit sphere in $\mathbb{R}^3$. That is,

$$S^2 = \{(x,y,z) \mid x^2 + y^2 + z^2 = 1\} .$$

For each P,Q in S^2, let d'(P,Q) be the length of the shortest path in S^2 from P to Q. Then $[S^2, d']$ is a metric space.

4. Let S be the set of all commercial airfields in North America. For each P,Q in S, let d(P,Q) be the minimum time (in hours) required to travel from P to Q by some combination of regularly scheduled flights. Then [S,d] is a metric space. Here we are assuming, contrary to two well-known facts, that every commercial plane flies exactly on schedule, and the flight-time from A to B is the same as the flight-time from B to A.

5. Let [X,d] be a metric space, and let d' = $\sqrt[3]{d}$. Then [X,d'] is a metric space.

6. Let F be the set of all continuous functions defined on the closed interval $[0,1] \subset \mathbb{R}$. For each f,g in F, let

$$d(f,g) = \int_0^1 |f(x) - g(x)| dx .$$

Then [F,d] is a metric space.

7. Let G be the set of all functions $[0,1] \to \mathbb{R}$ which are integrable in the sense of Riemann. For each f,g in G, let d(f,g) be as in Problem 6. Then [G,d] is a metric space. (Omit this problem if you don't know about functions which are integrable without necessarily being continuous.)

8. Let [X,d] be a metric space. Let d' be the function defined by the conditions

$$d'(P,Q) = d(P,Q) \quad \text{if} \quad d(P,Q) \leq 1 ,$$

$$d'(P,Q) = 1 \quad \text{if} \quad d(P,Q) > 1 .$$

Then [X,d'] is a metric space.

9. In a metric space [X,d], a set M is __bounded__ if the numbers d(P,Q) P,Q $\in$ M) form a bounded set. (That is, if there is a number b such that d(P,Q) $\leq$ b for each P,Q in M.) Let M_1, M_2, ... be a sequence of subsets of X, such that M_1 is bounded, and $M_{n+1} \subset M_n$ for each n. Then

$$\bigcap_{n=1}^{\infty} M_n \neq \emptyset.$$

10. Let F be as in Problem 6. For each f,g in F, let $d'(f,g)$ be the largest of the numbers $|f(x) - g(x)|$. Then $[F,d']$ is a metric space.

11. For $i = 1,2$, let a_i, b_i, c_i be non-negative real numbers. If (1) $a_i + b_i \geq o_i$, for $i = 1,2$, then (2)

$$\sqrt{a_1^2 + a_2^2} + \sqrt{b_1^2 + b_2^2} \geq \sqrt{c_1^2 + c_2^2} .$$

12. For $i = 1,2$, let a_i, b_i, c_i be non-negative real numbers. If (1') $a_i + b_i = c_i$, for $i = 1,2$, then (2) of Problem 11 holds.

13. Let $[X,d_1]$ and $[Y,d_2]$ be metric spaces. The product metric d for $X \times Y$ is defined by the formula

$$d((x_1,y_1),(x_2,y_2)) = \sqrt{d_1(x_1,x_2)^2 + d_2(y_1,y_2)^2} .$$

$[X \times Y, d]$ is a metric space.

14. An interval in $\mathbb{R}$ is a set of one of the forms

$$(a,b) = \{x | a < x < b\}, \ [a,b] = \{x | a \leq x \leq b\} ,$$

$$[a,b) = \{x | a \leq x < b\}, \ (a,b] = \{x | a < x \leq b\} ,$$

$$[a,\infty) = \{x | a \leq x\} , \qquad (-\infty,a] = \{x | x \leq a\},$$

$$(a,\infty) = \{x | a < x\} , \ (-\infty,a) = \{x | x < a\} , \ (-\infty,\infty) = \mathbb{R}.$$

Let f be a function $I \to \mathbb{R}$, where I is an interval. If $a < b \Rightarrow f(a) \leq f(b)$, for each a,b in I, then f is increasing. (If $a < b \Rightarrow f(a) < f(b)$, then f is strictly increasing. Similarly for decreasing and strictly decreasing.) Suppose that $a \geq 0$, and $I = [a,\infty)$, so that I is closed under addition; and let f be a function $I \to \mathbb{R}$. If $f(a + b) \leq f(a) + f(b)$, for every a,b in I, then f is sub-additive. State and prove a theorem which begins as follows: Theorem. Let f be a function $[0,\infty) \to [0,\infty)$, such that (1) $f(0) = 0$, (2) $f(x) > 0$ for every $x > 0$, (3) f is increasing, and (4) f is sub-additive. Then ... . Then show that the resulting theorem is a generalization of three true propositions which have been stated earlier.

Chapter 3: Neighborhood Spaces and Topological Spaces

Let G be a collection of sets, let G* be the union of the elements of G, and let X be a set. If $X \subseteq G^*$, then we say that G <u>covers</u> X.

Let [X,d] be a metric space. For each $P \in X$ and each $\varepsilon > 0$, let

$$N(P, \varepsilon) = \{Q \mid Q \in X \quad \text{and} \quad d(P,Q) < \varepsilon\} .$$

This set is called the ε-<u>neighborhood</u> of P (in X.) For example, in $\mathbb{R}^2$, with the Cartesian distance, every set $N(P, \varepsilon)$ is the interior of a circle, and conversely. Let $N = N(\mathbf{d})$ be the set of all such sets $N(P, \varepsilon)$. Then N has the following properties:

N.1. N covers X.

N.2. If $N_1, N_2 \in N$, and $P \in N_1 \subseteq N_2$, then there is an $N \in N$ such that

$$P \in N \subseteq N_1 \cap N_2 .$$

Here N.1 is trivial, but N.2 requires a proof. For this purpose, we need the following.

Theorem 1. In a metric space [X,d], let A and P be points, and let ε be a positive number, such that $P \in N(A, \varepsilon)$. Then there is a $\delta > 0$ such that $N(P, \delta) \subseteq N(A, \varepsilon)$.

From this we can get:

Theorem 2. For each metric space [X,d], $N(d)$ satisfies N.1 and N.2.

The collection $N = N(d)$ defined above is called the <u>neighborhood system</u> <u>induced by</u> d. More generally, if X is any non-empty set, and N is a collection of subsets of X, satisfying N.1 and N.2, then N is called a <u>neighborhood system</u> for X, and the pair $[X, N]$ is called a <u>neighborhood space</u>.

Let $[X, N]$ be a neighborhood space. A subset U of X is called <u>open</u> if for each point P of U there is an $N \in N$ such that $P \in N \subseteq U$.

Theorem 3. In a neighborhood space $[X, N]$, a set U is open if and only if U is the union of a collection of elements of N.

Theorem 4. Let $[X, N]$ be a neighborhood space, and let $\mathcal{O} = \mathcal{O}(N)$ be the set of all open sets in X. Then $\mathcal{O}$ has the following properties:

0.1. The empty set $\emptyset$ belongs to $\mathcal{O}$.

0.2. $X \in \mathcal{O}$.

0.3. The union of any collection of elements of $\mathcal{O}$ belongs to $\mathcal{O}$.

0.4. The intersection of any finite collection of elements of $\mathcal{O}$ belongs to $\mathcal{O}$.

The collection $\mathcal{O} = \mathcal{O}(N)$ is called the <u>topology induced by</u> the neighborhood system $\mathcal{N}$. In general, if X is any non-empty set, and $\mathcal{O}$ is a collection of subsets of X, satisfying 0.1-0.4, then $\mathcal{O}$ is called a <u>topology for</u> X, and the pair $[X,\mathcal{O}]$ is called a <u>topological space</u>. In $[X,\mathcal{O}]$, the elements of $\mathcal{O}$ are called <u>open</u> sets.

Evidently one possibility is that the only open sets (that is, the only elements of $\mathcal{O}$) are $\emptyset$ and X. Another possibility is that $\mathcal{O} = \mathcal{P}(X)$, where $\mathcal{P}(X)$ is the set of <u>all</u> subsets of X. $\mathcal{P}(X)$ is called the <u>discrete topology</u> for the set X, and a space of the type $[X, \mathcal{P}(X)]$ is called <u>discrete</u>.

It may easily happen that two different neighborhood systems for the same set X induce the same topology. For example, let X be the space $E = \mathbb{R}^2$; let $\mathcal{N}_1 = \mathcal{N}(d)$, and let $\mathcal{N}_2$ be the set of all interiors of squares in E. Then $\mathcal{O}(\mathcal{N}_1) = \mathcal{O}(\mathcal{N}_2)$. (Proof?) In general, if

$$\mathcal{O}(\mathcal{N}_1) = \mathcal{O}(\mathcal{N}_2) \ ,$$

then $\mathcal{N}_1$ and $\mathcal{N}_2$ are called <u>equivalent</u>. If d and d' are distance functions for the same set X, and

$$\mathcal{O}(\mathcal{N}(d)) = \mathcal{O}(\mathcal{N}(d')),$$

then d and d' are called <u>equivalent</u>.

Let $[X,\mathcal{O}]$ be a topological space. If there is a distance function d for X such that the given topology $\mathcal{O}$ is the same as $\mathcal{O}(\mathcal{N}(d))$, then $[X,\mathcal{O}]$ is <u>metrizable</u>. Thus a metrizable space is a topological space which <u>might</u> have been defined by means of a distance function. Note that a metrizable space is not the same thing as a metric space; in the latter, a particular distance function is given, as part of the structure. Note also that if the distance functions d_1 and d_2 are equivalent but different, then the metrizable spaces

$$[X, \ \mathcal{O}(\mathcal{N}(d_1))], \ \ [X, \ \mathcal{O}(\mathcal{N}(d_2))]$$

are exactly the same, although the metric spaces $[X,d_1]$ and $[X,d_2]$ are different.

In a topological space $[X,\mathcal{O}]$, let $P \in X$ and let M be a subset of X. If every open set (that is, every element of $\mathcal{O}$) that contains P contains a point of $M - P$, then P is a <u>limit point</u> of M. The union of M and the set of all limit points of M is called the <u>closure</u> of M, and is denoted by $\overline{M}$, or by $\mathrm{Cl}(M)$. If M contains all its limit points, then M is <u>closed</u>. The set of all limit points of M will be denoted by M^L. (The old standard notation is M', but this leads to trouble, because we need primes for other purposes.) Thus $\overline{M} = \mathrm{Cl}(M) = M \cup M^L$.

Let $[X,\mathcal{O}]$ be a topological space, and suppose that for every two (different) points of X there are disjoint open sets U_P and U_Q, containing P and Q respectively. Then $[X,\mathcal{O}]$ is called a <u>Hausdorff space</u>.

Given a topological space $[X,\mathcal{O}]$, and a non-empty subset M of X, let

$$\mathcal{O}|M = \{V|V = M \cap U \text{ for some } U \in \mathcal{O}\}.$$

Theorem 5. $\mathcal{O}|M$ is a topology for M.

Thus, given a topological space $[X,\mathcal{O}]$, $M \subset X$, $M \neq \emptyset$, the pair $[M,\mathcal{O}|M]$ is a topological space. This space is called a <u>subspace</u> of $[X,\mathcal{O}]$, and $\mathcal{O}|M$ is called the <u>subspace topology</u> for M. When subsets of $\mathbb{R}^n$ are regarded as topological spaces, the subspace topology will always be meant, unless some other topology is specifically mentioned.

In each of the problems below, the proposition stated is supposed to hold in every topological space, unless some additional hypothesis is stated. By a set we shall always mean a set of points (that is, a subset of X), unless the contrary is clear.

Problem Set 3

Prove or disprove:

1. Let $[X,\mathcal{O}]$ be a topological space. Then there is a neighborhood system $\mathcal{N}$ for X such that $\mathcal{O} = \mathcal{O}(\mathcal{N})$.

2. Let $[X,\mathcal{N}]$ be a neighborhood space. Then $\mathcal{N} \subset \mathcal{O}(\mathcal{N})$. That is, every neighborhood is open.

3. If U is open, and $M = X - U$, then M is closed.

4. If M is closed, and $U = X - M$, then U is open.

5. For every $M \subset X$, $\overline{M}$ is closed.

6. No set is both open and closed.

7. Every set is either open or closed.

8. Let A and B be sets (of points.) Then $(A \cup B)^L \subset A^L \cup B^L$. (That is, every limit point of $A \cup B$ is either a limit point of A or a limit point of B.)

9. For all sets A and B, $(A \cup B)^L = A^L \cup B^L$.

10. For all sets A and B, $\overline{A \cup B} = \overline{A} \cup \overline{B}$.

11. Every finite set is closed.

12. If M_1, M_2, ... are closed, then $\bigcap_{i=1}^{\infty} M_i$ is closed.

13. Let G be a collection of closed sets. Then $\bigcap_{g \in G} g$ is closed. (Is this more general than the proposition stated in Problem 12?)

14. If M_1 and M_2 are closed, then $M_1 \cup M_2$ is closed.

15. If M_1, M_2, ..., M_n are closed, then $\bigcup_{i=1}^{n} M_i$ is closed.

3. Neighborhood Spaces and Topological Spaces

16. If M_1, M_2, ... are closed, then $\bigcup\limits_{i=1}^{\infty} M_i$ is closed.

17. $P \in \bar{M}$ if and only if every open set that contains P contains a point of M.

18. Let X be any non-empty set, and let

$$\mathcal{O} = \{U \mid U \subset X \text{ and } U = \emptyset \text{ or } X - U \text{ is finite}\}.$$

Then $\mathcal{O}$ is a topology for X.

19. Every metrizable space is Hausdorff.

20. Every Hausdorff space is metrizable.

21. Let $[X,\mathcal{O}]$ be as in Problem 18. Then $[X,\mathcal{O}]$ is Hausdorff.

22. Let $[X,\mathcal{O}]$ be as in Problem 18. Then $[X,\mathcal{O}]$ is metrizable.

23. Every discrete space is metrizable.

24. Let $[X,d]$ be a metric space, and let $d' = \sqrt{d}$. Then $\mathcal{N}(d') = \mathcal{N}(d)$. (Thus, automatically, $\mathcal{O}(\mathcal{N}(d')) = \mathcal{O}(\mathcal{N}(d))$, and d and d' are equivalent.)

25. Let X, d, and d' be as in Problem 2.8. Then d and d' are equivalent.

26. Let $[X,d]$ be a metric space, let k be a positive number, and let

$$\mathcal{M} = \{N(P,\epsilon) \mid \epsilon < k\} \ .$$

Then $\mathcal{M}$ is a neighborhood system for X, and is equivalent to $\mathcal{N}(d)$.

27. In a topological space, the statement

$$\text{Lim}_{n \to \infty} P_n = P$$

means that for every open set U containing P there is an integer k such that

$$n \geq k \implies P_n \in U.$$

We then say that the sequence P_1, P_2, ... is <u>convergent</u>, and that it <u>converges to</u> P.

No sequence P_1, P_2, ... converges to each of two (different) points.

28. For every sequence of points, there is a point to which the sequence does <u>not</u> converge.

29. If $P \in M^L$, then there is a sequence P_1, P_2, ... of points of M, converging to P.

30. In a metrizable space, the proposition stated in Problem 29 holds true.

31. There is a topological space $[X,\mathcal{O}]$ such that (1) $X^L = X$ but (2) every convergent sequence is ultimately constant. Here (2) means that if $\text{Lim}_{n \to \infty} P_n = P$, then there is an integer k such that

$$n \geq k \implies P_n = P.$$

(Note that (1) rules out discrete spaces, in which (2) is satisfied trivially.)

    32.   Every subspace of a Hausdorff space is Hausdorff.

    33.   Every subspace of a metrizable space is metrizable.

    34.   For every set M, $Cl(\overline{M}) = \overline{M}$.

    35.   If U is open, and $P \in U$, then there is an open set V such that $P \in V$ and $\overline{V} \subset U$.

    36.   The proposition stated in Problem 35 holds in every metrizable space.

Chapter 4: Cardinality. Finite and Countable Sets

In the preceding sections, we have been making informal use of the idea of a finite set. This idea is worth re-examining, and also we need to know about various types of infinite sets.

Let A and B be non-empty sets. If there is a bijection f: A ⟷ B, then A and B are called <u>cardinally equivalent</u> (or simply <u>equivalent</u>,) and we write

$$A \sim B.$$

<u>Theorem 1</u>. Let A, B, and C be non-empty sets. Then

(1) A ~ A.

(2) A ~ B ⟹ B ~ A, and

(3) A ~ B and B ~ C ⟹ A ~ C.

(We might have said, more briefly, that ~ is an equivalence relation. But the consequences of this innocent-looking statement would have been logically catastrophic. To be exact, a binary relation on a set S is a subset R of S × S; when we write aRb (a stands in the relation R to b,) this means that (a,b)∈ R. If ~ is regarded as a relation, then it must be a subset of S × S, where S is the set of all sets (!?!), which constitutes one of the paradoxes of various naive formulations of mathematical logic.)

We recall that $\mathbb{Z}^+$ is the set of all positive integers; and for each n in $\mathbb{Z}^+$,

$$I_n = \{1, 2,...,n\} = \{i \mid i \in \mathbb{Z}^+ \text{ and } 1 \leq i \leq n\} .$$

Such sets are called <u>segments</u> of $\mathbb{Z}^+$.

There are now two ways of defining the terms <u>finite</u> and <u>infinite</u>, as follows.

<u>Definition</u>. A set A is <u>finite</u> if (1) A = ∅ or (2) A ~ I_n for some n. A set is <u>infinite</u> if it is not finite.

<u>Definition</u>. A set A is <u>infinite in the sense of Dedekind</u> (briefly, <u>infinite-D</u>) if there is a bijection between A and a proper subset of A. A set is <u>finite-D</u> if it is not infinite-D.

It will turn out that these definitions are equivalent, and so the phrase "in the sense of Dedekind" will eventually become superflous. But for the purposes of this section, the definition using the sets I_n is much more convenient as a working definition; the point is that the <u>existence</u> of a bijection A ⟷ I_n is a more easily usable hypothesis than the <u>non</u>-existence of a bijection between A and a proper subset of A.

Topology

Here and hereafter, we shall assume that the following are known.

(1) For each n in $\mathbb{Z}^+$, n+1 is the smallest integer that is greater than n.

(2) For each n in $\mathbb{Z}^+$, $I_n \cup \{n+1\}$.

These require proofs, but the proofs would take us back to the foundations of analysis, and every course has got to start somewhere. We shall, however, need to make explicit use of the induction principle, in one of the following forms.

The Induction Principle (First Form.) Let S be a set of positive integers. If (1) $1 \in S$ and (2) for each n, $n \in S \Rightarrow n+1 \in S$, then (3) $S = \mathbb{Z}^+$.

Induction Principle. (Second Form.) Let P_1, P_2, ... be a sequence of propositions. If (1) P_1 is true and (2) for each n, $P_n \Rightarrow P_{n+1}$, then (3) each of the propositions P_n is true.

Theorem 2. If A is finite and B is a singleton, then $A \cup B$ is finite.

(We recall that a singleton is a set of the form B = {b}. That is, B is a singleton if there is a $b \in B$ such that $x \in B \Rightarrow x = b$.

Theorem 3. Let A be a set, and let $a_0 \in A$. Let f: $A \longleftrightarrow B$ be a bijection between A and a proper subset B of A. Then there is a bijection g: $A \longleftrightarrow B'$, between A and a proper subset B' of A, such that $g(a_0) = a_0$.

Theorem 4. If A is finite-D, and B is a singleton, then $A \cup B$ is finite-D.

Theorem 5. Every set I_n is finite-D.

Theorem 6. If A is finite, then A is finite-D.

Theorem 7. If $A \sim I_m$ and $A \sim I_n$, then m = n.

If $A \sim I_n$, then we write

$$CardA = n.$$

We also write

$$Card\emptyset = 0.$$

Note that by Theorem 7, the suggestion conveyed by the "functional" notation is correct: every finite set determines a unique integer $n \geq 0$. This integer n is called the number of elements of A, or the cardinal number of A.

Theorem 8. For each $n \in \mathbb{Z}^+$, every subset of I_n is finite.

Theorem 9. Every subset of a finite set is finite.

Theorem 10. The union of any two finite sets is finite.

Theorem 11. If $A \subset B$, and A is finite and B is infinite, then B - A is infinite.

Theorem 12. The union of any finite collection of finite sets is finite.

Theorem 13. $\mathbb{Z}^+$ is infinite-D.

Theorem 14. $\mathbb{Z}^+$ is infinite.

A set A is countable if (1) $A = \emptyset$ or (2) there is an (infinite) sequence

$$a_1, a_2, \ldots$$

4. Cardinality. Finite and Countable Sets

of elements of A in which every element of A appears at least once, so that

$$A = \{a_1, a_2, \ldots \} \ .$$

Note that duplicates are allowed in the sequence, and so every finite set is countable: given $A = \{a_1, a_2, \ldots, a_n\}$, we can let $a_i = a_n$ for each $i \geq n$, and this gives $A = \{a_1, a_2, \ldots\}$. If $A = \{a_1, a_2, \ldots \}$, with $a_i \neq a_j$ for $i \neq j$, then $A \sim \mathbb{Z}^+$. In this case, we say that A is <u>countably infinite</u>, and we write

$$CardA = \aleph_0 \ .$$

<u>Theorem 15.</u> If A is countable, then either (1) A is finite or
(2) $CardA = \aleph_0$.

<u>Theorem 16.</u> The union of any countable collection of finite sets is countable.

<u>Theorem 17.</u> The union of any countable collection of countable sets is countable.

<u>Theorem 18.</u> Every subset of a countable set is countable.

Problem Set 4.

Prove or disprove:

1. Let $\mathbb{Z}$ be the set of all integers. Show directly (that is, without appealing to more general theorems,) that $\mathbb{Z} \sim \mathbb{Z}^+$.

2. Let $\mathbb{Q}$ be the set of all rational numbers. Then $\mathbb{Q} \cap [0,1]$ is countable.

3. $\mathbb{Q}$ is countable.

4. Let A be the set of all ordered pairs (a_1, a_2) of integers. Then A is countable.

5. Let B be the set of all finite sequences $a_1, a_2, \ldots, a_n$ of integers. Then B is countable.

6. Let C be the set of all (infinite) sequences of integers. Then C is countable.

7. A real number x is <u>algebraic</u> if it is a root of an equation of the form

$$a_n x^n + a_{n-1} x^{n-1} + \ldots + a_1 x + a_0 = 0,$$

where the coefficients a_i are integers and $a_n \neq 0$. Let D be the set of all algebraic numbers. Then D is countable.

8. A real number is <u>transcendental</u> if it is not algebraic. Let T be the set of all transcendental numbers. Then T is uncountable.

9. We recall that for each set A, $\mathcal{P}(A)$ is the set of all subsets of A. Let S be the set of all sequences of positive integers. Then $S \sim \mathcal{P}(\mathbb{Z}^+)$.

10. Let G be a collection of sets of positive integers. (That is, $G \subset \mathcal{P}(\mathbb{Z}^+)$.) Suppose that for each $g, g' \in G$, either $g \subset g'$ or $g' \subset g$. Then G is countable.

11. Let $E = \{i \mid i \in \mathbb{Z}^+ \text{ and } 1 \le i \le 8\}$, and let F be the set of all infinite sequences of elements of E. Then F is countable.

Chapter 5: The Completeness of IR. Uncountable Sets

We now approach the proof that $\mathbb{R}$ is uncountable. Since the set $\mathbb{Q}$ of all rational numbers is countable, it is clear that any valid proof of the uncountability of $\mathbb{R}$ must use the continuity of $\mathbb{R}$. The classical formulation of this is as follows.

An _ordering_ of a set S is a binary relation $<$, defined on S, satisfying the following conditions.

0.1. $x < x$ _never_.

0.2. For each $x,y,z \in S$, $x < y$ and $y < z \Longrightarrow x < z$.

If $<$ is an ordering of S, then the pair $[S,<]$ is called an _ordered set_. A _linear_ ordering of S is an ordering satisfying the following:

0.3. For each $x,y \in S$, at least one of the following holds:

$$x < y, \quad \text{or} \quad x = y, \quad \text{or} \quad y < x .$$

We then say that $[S,<]$ is a _linearly ordered set_. Thus $[\mathbb{R},<]$ is a linearly ordered set, when $<$ is the usual ordering of $\mathbb{R}$.

In a linearly ordered set $[S,<]$, consider a set $A \subset S$. If $b \in S$, and $a \leq b$, for each $a \in A$, then b is an _upper bound_ of A. If such a b exists, then A is _bounded above_. (Similarly for _lower bound_ and _bounded below_.) If every non-empty set $A \subset S$ which is bounded above has a least upper bound, then $[S,<]$ is _complete in the sense of Dedekind_ (briefly, _complete-D_.) The least upper bound of A is denoted by

$$\text{lub } A.$$

The Least Upper Bound Postulate (LUBP). $[\mathbb{R},<]$ is complete-D.

On this basis, the following can be proved.

Theorem 1. (The Nested Interval Postulate, NIP.) Let

$$[a_1,b_1], \ [a_2,b_2], \ \ldots$$

be a sequence of closed intervals in $\mathbb{R}$, and suppose that the sequence is _nested_, in the sense that

$$[a_{i+1},b_{i+1}] \subset [a_i,b_i]$$

for each i. Then

$$\bigcap_{i=1}^{\infty} [a_i,b_i] \neq \emptyset .$$

We call this NIP because it is sometimes taken as a postulate. It can be used to prove the following.

Theorem 2. $\mathbb{R}$ is uncountable.

You should not use Georg Cantor's "diagonal" proof unless you are prepared to make it logically complete by setting up a decimal system, first for the integers and then for the real numbers.

If $A \sim \mathbb{R}$, then we write

$$\text{Card} A = \underline{c}.$$

The infinite cardinal $\underline{c}$ is called the cardinal of the continuum. To avoid various logical problems, we may prefer to consider that the symbols "CardA" and "$\underline{c}$" have no meaning when they stand alone, and interpret the expression "CardA = $\underline{c}$" merely to mean that $A \sim \mathbb{R}$. Similarly for statements of the form $\text{Card} A = \aleph_0$.

Uncountable sets arise in simpler ways, as follows.

Theorem 3. $\mathcal{P}(\mathbb{Z}^+)$ is uncountable.

The proof is indirect. More generally:

Theorem 4. (Georg Cantor.) Let Z be any set. Then Z and $\mathcal{P}(Z)$ are not cardinally equivalent. In fact, no function

$$f: Z \longrightarrow \mathcal{P}(Z)$$

is surjective.

If A and B are not cardinally equivalent, then we write $A \not\sim B$. Thus $Z \not\sim \mathcal{P}(Z)$ every Z. Note that the validity of this statement begins with the empty set, which has one subset, but no elements.

Problem Set 5

1. For each $a < b$, $c < d$ in $\mathbb{R}$, $[a,b] \sim [c,d]$.
2. $[0,1] \sim [0,1)$.
3. $[0,1] \sim (0,1)$.
4. $(-\pi/2, \ \pi/2) \sim \mathbb{R}$.
5. $\text{Card}(0,1) = \underline{c}$.
6. No interval in $\mathbb{R}$ is countable.
7. $\mathbb{R}^2$ is uncountable.
8. If $A \subset \mathbb{R}$, and A contains an open interval, then $A \sim \mathbb{R}$.
9. If $A \subset \mathbb{R}$, and $A \sim \mathbb{R}$, then A contains an open interval.
10. $[0,1) \times [0,1) \sim [0,1)$.

_____

Now that you know that some sets are uncountable, you may find that some problems in earlier sections have become easier.

Chapter 6: The Schröder-Bernstein Theorem

Let A and B be sets. If A ~ B' $\subseteq$ B, for some B', then we write A $\leq$ B. If A $\leq$ B, but A $\not\sim$ B, then we write A < B.

(The notation < conveys several suggestions, only one of which is obviously true: by definition of <, A < A is impossible.)

Theorem 1. If A $\subseteq$ B $\subseteq$ C, and A ~ C, then A ~ B.

Theorem 2. (The Schröder-Bernstein Theorem.) Let A and B be sets. If A $\leq$ B and B $\leq$ A, then A ~ B.

Theorem 3. If A < B $\leq$ C, or A $\leq$ B < C, then A < C.

It is now natural to ask whether any two sets are comparable, in the sense of cardinality. That is, do we always have A < B, or A ~ B, or B < A? It turns out that this is true, but the proof is _much_ harder than those of the preceding theorems. So far, we have been manipulating and combining bijections which were given by hypothesis. To prove the existence of an injection A $\longrightarrow$ B or B $\longrightarrow$ A, when no function at all is given by hypothesis, is a task of a higher order of difficulty.

<u>Problem Set 6</u>

Re-examine the problems in the preceding sections, to see which of them are made easy by the Schröder-Bernstein Theorem. Then consider the following.

1. Let A be a set which is infinite-D. Then A contains a set which is countably infinite.

2. Let G be a collection of sets of positive integers. Suppose that for each g, g' $\in$ G, either g $\subseteq$ g' or g' $\subseteq$ g. Then G is countable. (You have seen this one before.)

3. It is a fact that every infinite set contains a countably infinite set. Try to find a way to convince yourself of this, not by rhetoric but by a line of reasoning in which the gaps might conceivably be filled. At the present stage, it is improbable in the last degree that you will be able to fill the gaps, but the enterprise proposed here is nevertheless worthwhile. Note the difference between this problem and Problem 1: in the latter, a bijection is given by hypothesis.

4. A set M $\subseteq$ $\mathbb{R}$ has the <u>Heine-Borel Property</u> if for each collection G of open intervals, covering M, there is a finite subcollection G' of G which also covers M. <u>Theorem (?)</u>. If M has the Heine-Borel property, then M is closed and bounded.

5. For $1 \leq i < n$, let M_i be a subset of $\mathbb{R}$. If each of the sets M_i has the Heine-Borel Property, then their union has the same property.

Chapter 7: Compactness in $\mathbb{R}^n$

In the theory of functions of one real variable, the following is fundamental.

Theorem 1. (The Heine–Borel Theorem.) Let G be a collection of open intervals, covering the closed interval $[a,b]$. Then some finite subcollection of G also covers $[a,b]$.

Such infinite coverings arise naturally. Suppose, for example, that f is a function $[a,b] \rightarrow \mathbb{R}$. If f is continuous at a point x_0 of $[a,b]$, then for every $\epsilon > 0$ there is a $\delta_{x_0} > 0$ such that

$$x \in [a,b] \text{ and } |x - x_0| < \delta_{x_0} \implies |f(x) - f(x_0)| < \epsilon .$$

(In some treatments, this is the definition of continuity, and in others it is an easy theorem.) If f is continuous at every point x of $[a,b]$, then for each $x \in [a,b]$ there is a $\delta_x > 0$ such that

$$x' \in [a,b] \text{ and } |x - x'| < \delta_x \implies |f(x') - f(x)| < \epsilon .$$

Thus we have an infinite collection

$$G = \{(x - \delta_x, x + \delta_x)\} ,$$

covering $[a,b]$. By the Heine–Borel Theorem, some finite subcollection also covers.

Theorem 2. Let f be a continuous function $[a,b] \rightarrow \mathbb{R}$. Then f is bounded.

(This is given merely as a simple example of what the Heine–Borel Theorem is good for in analysis.)

Example 1. The Heine–Borel Theorem depends essentially on the completeness of $\mathbb{R}$. Consider what happens if we replace $\mathbb{R}$ by the set $\mathbb{Q}$ of all rational numbers. For each $a, b \in \mathbb{Q}$, we define the "closed interval" $[a,b]'$ to be $[a,b] \cap \mathbb{Q}$; and similarly the "open interval" $(a,b)'$ is $(a,b) \cap \mathbb{Q}$. There is a collection G of "open intervals", covering $[0,1]'$, such that no finite subcollection of G covers $[0,1]'$.

Example 2. When $\mathbb{R}$ is replaced by $\mathbb{Q}$, it is not just the proof of Theorem 2 that breaks down; the theorem itself becomes false. There is a continuous function $f \colon [0,1]' \rightarrow \mathbb{Q}$, such that f is unbounded.

The following generalization of the Heine–Borel Theorem is easy.

Theorem 3. Let $[a,b]$ be a closed interval in $\mathbb{R}$, and let H be a collection of open sets, covering $[a,b]$. Then some finite subcollection of H covers $[a,b]$.

In abstract spaces, the conclusion of Theorem 3 is used as the definition of compactness. To be exact:

7. Compactness in $\mathbb{R}^n$

Definition. Let M be a set of points, in a topological space $[X, \mathcal{O}]$. If every collection of open sets covering M contains a finite subcollection which covers M, then M is <u>compact</u>.

In this language, we can restate Theorem 3:

Theorem 3'. In $\mathbb{R}$, every closed interval is compact.

A set M in $\mathbb{R}$ is <u>bounded</u> if it lies on some closed interval $[a,b]$. (If so, M lies on some closed interval of the type $[-k,k]$, and we have $|x| \leq k$ for each $x \in \mathbb{R}$. The converse is equally trivial.) Theorem 3' has the following generalization:

Theorem 4. In $\mathbb{R}$, every closed bounded set is compact.

In much the same way, NIP can be generalized:

Theorem 5. Let M_1, M_2, ... be a sequence of non-empty closed sets in $\mathbb{R}$, such that (1) M_1 is bounded and (2) for each i, $M_{i+1} \subset M_i$. Then

$$\bigcap_{i=1}^{\infty} M_i \neq \emptyset .$$

A <u>closed interval</u> in $\mathbb{R}^n$ is a set of the form

$$J = \{(x_1, x_2, \ldots, x_n) \mid a_i \leq x_i \leq b_i \quad \text{for} \quad 1 \leq i \leq n\},$$

where the a_i's and b_i's are constants, and $a_i < b_i$ for each i.

Theorem 6. If J_1, J_2, ... is a nested sequence of closed intervals in $\mathbb{R}_n$, then $\bigcap_{i=1}^{\infty} J_i \neq \emptyset$.

Theorem 7. (The Heine-Borel Theorem.) Let J be a closed interval in $\mathbb{R}^n$, and let G be a collection of open intervals covering J. Then some finite subcollection of G covers J.

(Definition of open intervals in $\mathbb{R}^n$?)

Theorem 8. In $\mathbb{R}^n$, every closed interval is compact.

Theorem 9. Every closed and bounded set in $\mathbb{R}^n$ is compact.

Theorem 10. Let M_1, M_2, ... be a sequence of non-empty closed sets in $\mathbb{R}^n$, such that (1) M_1 is bounded and (2) for each i, $M_{i+1} \subset M_i$. Then

$$\bigcap_{i=1}^{\infty} M_i \neq \emptyset .$$

Theorem 11. (The Bolzano-Weierstrass Theorem). In $\mathbb{R}$ (or $\mathbb{R}^n$), every bounded infinite set has a limit point.

Consider the following recursive scheme for defining a subset C of $[0,1]$.

(1) Let $C_0 = [0,1]$.

(2) Suppose that we have given a set $C_i \subset [0,1]$, such that C_i is the union of 2^i disjoint closed intervals $[a,d]$. For each such $[a,d]$ we choose points

b and c such that $a < b < (a+d)/2 < c < d$. In C_i, we replace each $[a,d]$ by $[a,b] \cup [c,d] = [a,d] - (b,c)$. This gives a set C_{i+1} which is the union of 2^{i+1} disjoint closed intervals. Let

$$C = \bigcap_{i=1}^{\infty} C_i .$$

Such a set C is called a <u>Cantor set</u>.

The classical example is the "middle–third" Cantor set. To obtain this, we always use the "open middle third" of $[a,d]$ as the interval (b,c). That is, we use $b = a + (d-a)/3$, $c = d - (d-a)/3$. We have avoided this special definition, because the resulting set is capable of being investigated by methods which have little to do with the concepts of topology. And the extra generality is significant in other connections. See Problem 11 below.

Problem Set 7.

Investigate the following propositions. Throughout, it should be understood that C is a Cantor set, but not necessarily the middle–third Cantor set.

1. C is closed.

2. Every point of C is a limit point of C.

3. Every point of C is an end point of one of the intervals (b,c) that we deleted, in passing from some set C_i to the next set C_{i+1}.

4. C is countable.

5. Let S be the set of all (infinite) sequences of 0's and 1's. Then S is uncountable.

6. Let T be the set of all sequences in S which are ultimately constant and $=1$. That is, x_1, x_2, ... belongs to T if there is an N such that for each $i \geq N$, $x_i = 1$. Then T is countable.

7. $S \sim S - T$.

8. Card$S = \underline{c}$.

9. C contains no interval.

Let K be a closed bounded set of real numbers. Let $G = \{(a_i, b_i)$ be a finite collection of open intervals, covering K. For each such G, let

$$L(G) = \sum (b_i - a_i);$$

and let

$$mK = glb \{L(G)\} .$$

mK is called the <u>measure</u> of K. If $mK = 0$, then M is <u>null</u>.

10. <u>Theorem</u>. $m[a,b] = b - a$.

11. <u>Theorem</u>. For each α $[0,1)$ there is a Cantor set C such that $mC = \alpha$.

74

12. Let C be a Cantor set. Then $C \sim C \times C$.

13. Theorem (?). Let M_1, M_2, ... be sets of real numbers. If for each i, M_i is null, then

$$\bigcup_{i=1}^{\infty} M_i$$

is null.

Chapter 8: Compactness in Abstract Spaces

Theorem 7.9 asserts that in $\mathbb{R}^n$, every closed and bounded set is compact. The converse is also true.

Theorem 1. In $\mathbb{R}^n$, every compact set is closed and bounded.

But no such theorem can hold for topological spaces in general, where the idea of boundedness is not defined. And boundedness, in an abstract metric space, has no topological meaning at all: in Problem 3.25, we found that every distance function is equivalent to a bounded distance function. Given a metric space $[X,d]$, we let

$$d'(P,Q) = \begin{cases} d(P,Q) & \text{if } d(P,Q) \le 1, \\ 1 & \text{if } d(P,Q) > 1. \end{cases}$$

Then d' is a distance function, and is equivalent to d.

Therefore new ideas are needed, for the investigation of compactness in abstract spaces. We can, however, generalize half of Theorem 1, as follows.

Theorem 2. In a Hausdorff space, every compact set is closed.

We found (Theorem 7.11, the Bolzano-Weirstrass Theorem) that in $\mathbb{R}^n$, every bounded infinite set has a limit point. This idea is closely related to compactness.

Definition. Let M be a set in a topological space. If every infinite set in M has a limit point which belongs to M, then M is BW-compact.

Theorem 3. In any topological space, every compact set is BW-compact.

A space $[X,\mathcal{O}]$ is separable if there is a countable set $S \subset X$ such that $X = \bar{S}$. If there is a countable collection $\mathcal{N}$ of open sets such that $\mathcal{N}$ is a neighborhood system for X and $\mathcal{O} = \mathcal{O}(\mathcal{N})$, then $[X,\mathcal{O}]$ is completely separable. (If $\mathcal{O} = \mathcal{O}(\mathcal{N})$, then $\mathcal{N}$ is called a basis for $\mathcal{O}$. Thus $[X,\mathcal{O}]$ is completely separable if $\mathcal{O}$ has a countable basis.)

Theorem 4. Every completely separable space is separable.

Theorem 5. Every separable metrizable space is completely separable.

Theorem 6. If $[X,\mathcal{O}]$ is metrizable, and is BW-compact, then $[X,\mathcal{O}]$ is separable (and hence completely separable).

Theorem 7. In a metrizable space, every BW-compact set is compact.

For Hausdorff spaces in general, this theorem is false. But the search for a counter-example would be unreasonably difficult at this stage.

Theorem 8. In a metrizable space, let M_1, M_2, ... be a nested sequence of non-empty compact sets. Then

$$\bigcap_{i=1}^{\infty} M_i \ne \emptyset .$$

Theorem 9. In a metrizable space, let M be a compact set such that $M = M^L$.
Then M is uncountable.

Note that the uncountability of $\mathbb{R}$ is a corollary of this theorem.

In a topological space, P is a condensation point of M if every open set
that contains P contains an uncountable subset of M.

Let a_1, a_2, ... be a sequence, and let n_1, n_2, ... be a strictly increasing
sequence of positive integers. For each i, let $b_i = a_n$. Then b_1, b_2, ... is
a subsequence of a_1, a_2, ...

Problem Set 8

Investigate the following.

1. In a separable space, every collection of disjoint open sets is countable.

2. Every separable space is completely separable.

3. Let $[X,\mathcal{O}]$ be a topological space. If every collection of disjoint open
sets is countable, then $[X,\mathcal{O}]$ is completely separable.

4. In a separable space, every uncountable set of points has a condensation
point.

5. In a completely separable space, every uncountable set has a condensation
point.

6. In $\mathbb{R}$ there is an uncountable set with one and only one condensation point.

7. In a topological space, a set S is non-dense if $\bar{S}$ contains no open set.
Theorem (?). No compact metrizable space is the union of a countable collection of
non-dense sets.

8. Let P_1, P_2, ... be a sequence of points, in a metric space [S,d].
Suppose that for every $\varepsilon > 0$ there is an $N \in \mathbb{Z}^+$ such that

$$n, \ m \geq N \implies d(P_n, P_m) < \varepsilon \ .$$

Then the sequence P_1, P_2, ... is regular. If every regular sequence is convergent,
then [X,d] is complete. Theorem (?). No complete metric space is the union of
a countable collection of non-dense sets.

9. Theorem. In a metric space [X,d], let M be a compact set, and let G
be an open covering of M. Then there is an $\varepsilon > 0$ such that for each point P
of M, $N(P,\varepsilon)$ lies in some one element g of G. The least upper bound of these
numbers α is called the Lebesgue number of G (with respect to M).

In a metric space, the diameter δK of a bounded set K is lub$\{d(P,Q)|P,Q \in K\}$.
If G is an open covering of a compact set M, then by the result of Problem 9 it
follows that there is an $\alpha > 0$ such that if $K \subset M$, and $\delta K < \alpha$, then K lies in
a single element g of G.

10. In a BW-compact metric space, every sequence of points has a convergent
subsequence.

Chapter 9: The Use of Choice in Existence Proofs

In this section we shall give full discussions and furnish some proofs, because we shall be dealing not with the substance of topology but with various fine points in what one might call Applied Mathematical Logic.

The methods that you must have been using so far, in proving the existence of certain sequences, involve curious logical points which deserve careful examination, partly in order to make the proofs complete and partly because the devices used to make the proofs complete will be technically useful later. To avoid presenting the solution of one of the problems above, we shall illustrate one of these points with a proof which is at least slightly new.

<u>Theorem A</u>. Let a_1, a_2, ... be a sequence of positive real numbers, such that 0 is a limit point of the set $A = \{a_i\}$. Then there is a strictly decreasing sequence b_1, b_2, ... of points of A such that $\lim_{n \to \infty} b_n = 0$.

"Proof. (1) Let j be the least integer such that $a_j < 1$, and let $b_1 = a_j$."

"(2) Suppose that b_i has been defined for $i \le n$. Let k be the least integer such that $a_k < b_n$ and $a_k < 1/(n+1)$, and let $b_{n+1} = a_k$."

Here it appears that we have "defined a sequence by induction". But the "proof" is not a straightforward application of the Induction Principle. If it were, then we must have been letting

$$S = \{n \mid b_i \text{ has been defined for } i \le n\}.$$

But the condition in the braces raises a question: <u>Who</u> has defined b_i, for $i \le n$, and when did he do it? The set S seems to vary as time passes; S seems to depend on how many times the unknown person has gone through the procedure described in (2).

Nevertheless, proofs like the above are standard, for the good reason that they are essentially valid. And the task of showing that they are valid is a straightforward chore, as follows.

We recall that a <u>sequence</u> is a function

$$f: \mathbb{Z}^+ \longrightarrow B, \qquad i \longmapsto f(i) = b_i,$$

where B is any set. A <u>finite sequence</u>, <u>of length</u> n, is a function

$$f_n: I_n \longrightarrow B,$$

where $I_n = \{1, 2, \ldots, n\}$. If f is a sequence, and

9. The Use of Choice in Existence Proofs

$$f_n = f|I_n : I_n \longrightarrow B,$$

then f_n is a <u>segment</u> of f; and similarly, if f_n is a finite sequence of length
n, and m $\leq$ n, then the function

$$f_m = f_n|I_m : I_m \longrightarrow B$$

is called a <u>segment</u> of f_n.

Let F be a collection of functions. Then F* denotes the union of the
elements of F. This is a notation that we have been using all along, because each
f $\in$ F is a collection of ordered pairs. There are obvious examples to show that
F* is not necessarily a function.

<u>Theorem 1</u>. (The Unique Finite Sequences Theorem, UFST.) Let A be a non-
empty set, and let F be a collection of finite sequences of the type $f_n : I_n \longrightarrow A$.
Suppose that (1) for each n, F contains one and only one finite sequence of
length n and (2) if $f_n \in$ F, then every segment of f_n belongs to F. Then F*
is a sequence f, and every element of F is a segment of f.

Once this theorem has been stated, its proof consists merely of easy verifi-
cations. And it is enough to rescue the proof of Theorem A. We let F be the
set of all <u>finite</u> sequences formed according to rules (1) and (2). Every two such
sequences agree, as far as they go, and so F satisfies the condition of UFST.
Now F* is the desired sequence b_1, b_2, ...

In normal practice, it would be tedious and needless to formalize proofs of
this kind by explicit use of UFST. It is important, however, to understand that
such proofs <u>can</u> be formalized.

In the problem above, a sequence a_1, a_2, ... was given, and we used it in
defining the collection F. Often, however, we need to prove the existence of a
sequence in cases where no sequence at all is given by hypothesis; and in many such
cases we need the following apparatus.

<u>The Axiom of Choice</u>. Let G be a non-empty collection of disjoint non-empty
sets. Then there is a subset Z of G* such that for each g $\in$ G, Z $\cap$ g has
exactly one element.

Rarely can this be applied directly. Ordinarily, we need an easy theorem
based on it, as follows.

<u>Theorem 2</u>. (The Choice Function Theorem, CFT.) Let H be any non-empty
collection of non-empty sets (disjoint or not). Then there is a function

$$\phi : H \longrightarrow H^* ,$$

such that for each h $\in$ H,

$$\phi(h) \in h .$$

Proof. For each $h \in H$, let $g_h = \{(h,x) | x \in h\}$. Then the sets g_h are disjoint. Let $G = \{g_h | h \in H\}$, and let ϕ be the set Z given by the Axiom of Choice. Then ϕ is a function of the sort that we wanted.

In many cases, CFT is easier to apply than UFST. For example:

<u>Theorem B</u>. In a metric space $[X,d]$, if the point P is a limit point of the set M, then there is a sequence P_1, P_2, ... of points of M, converging to P.

Proof. Let H be the set of all non-empty subsets of X, and let

$$\phi: H \longrightarrow H^* = X$$

be as in CFT. For each n, let

$$M_n = N(P, 1/n) \cap (M - P).$$

Then $M_n \neq 0$ for each n. For each n, let $P_n = \phi(M_n)$. Then for each positive integer k, $n \geq k \Rightarrow P_n \cap N(P, 1/k)$. Therefore P_1, P_2, ... converges to P.

In this case, the points P_n could be chosen independently of one another, and so a single application of CFT was enough. But sometimes we need to use UFST and CFT in combination. For example, the natural proof of Theorem 5.2 can be formalized as follows.

Suppose that $\mathbb{R}$ is countable, $= \{x_1, x_2, \dots \}$. Let H be the set of all non-empty collections of closed intervals in $\mathbb{R}$. Let ϕ be a function $H - H^*$, as in CFT. Let F be the set of all finite sequences J_1, J_2, $\dots$, J_n of closed intervals such that:

(1) $J_1 = \phi(h_1)$, where h_1 is the set of all closed intervals J such that $x_1 \not\subset J$, and

(2) For each $i < n$, $J_{i+1} = \phi(h_{i+1})$, where h_{i+1} is the set of all closed intervals J such that $J \subset J_i$ and $x_{i+1} \not\subset J$.

Then F satisfies the conditions of UFST. Let f be the sequence given by UFST, and for each i, let $J_i = f(i)$. By NIP (Theorem 5.1), $\cap J_i \neq \emptyset$. But $\cap J_i$ must be empty, because $\mathbb{R} = \{x_1, x_2, \dots\}$, and for each k, $x_k \not\subset J_k$.

There is at least one ad hoc device which makes this proof work without any appeal to CFT. It is recommended that you hunt for such a device only if for some reason you find the enterprise amusing.

We have remarked that formal citations of UFST are hardly necessary. But CFT is another matter. When you need it, you should set up the necessary apparatus and cite it explicitly.

Chapter 10: Linearly Ordered Spaces

Let $[X,<]$ be a linearly ordered set, in the sense defined in Section 5. We define the induced neighborhood system $N = N(<)$ in the following way. For each $a \in X$, let

$$(-\infty,a) = \{x | x < a\}, \qquad (a,\infty) = \{x | a < x\} .$$

Let $M(<)$ be the set of all such "infinite intervals", and let $N(<)$ be the set of all non-empty subsets N of X such that (1) $N = X$ or (2) N is the intersection of a finite number of elements of $M(<)$. Under these conditions, the pair $[X,N(<)]$ is evidently a neighborhood space. It is not hard to see that the neighborhoods N are the non-empty sets of the types

$$(-\infty,\infty) = X, \quad (-\infty,a), \quad (a,\infty),$$

and

$$(a,b) = \{x | a < x < b\} .$$

The topology $\mathcal{O}(N(<))$ is called the topology induced by $<$, and is denoted by $\mathcal{O}(<)$. Let $\mathcal{O}$ be a topology for X. If $\mathcal{O} = \mathcal{O}(<)$, for some linear order relation $<$, then $[X,\mathcal{O}]$ is called a linearly ordered space.

It is not hard to see that $[\mathbb{R}, \mathcal{O}(N(d))]$ is a linearly ordered space, because $\mathcal{O}(N(d)) = \mathcal{O}(<)$, where $<$ is defined as usual. But the above definitions apply to much more general situations.

Given two linearly ordered sets $[X,<_1]$ and $[Y,<_2]$, the lexicographic ordering of the product $X \times Y$ is defined by the condition that $(x,y) < (x',y')$ if (1) $x <_1 x'$ or (2) $x = x'$ and $y <_2 y'$. The relation $<$ is called the lexicographic product of $<_1$ and $<_2$.

Theorem 1. If $<_1$ and $<_2$ are linear orderings of X and Y respectively, then their lexicographic product is a linear ordering of $X \times Y$.

Theorem 2. Let $[X,<_1]$ and $[Y,<_2]$ be linearly ordered spaces, each of which is complete-D, and let $<$ be the lexicographic product of $<_1$ and $<_2$. If Y has both a greatest and a least element, then $[X \times Y,<]$ is complete-D.

In any linearly ordered set, closed intervals are defined in the same way as in $\mathbb{R}$. That is,

$$[a,b] = \{x | a \leq x \leq b\} .$$

Theorem 3. If $[X,<]$ is complete-D, then every closed interval in $[X,<]$ is compact.

Problem Set 10.

Prove or disprove.

1. In a linearly ordered space, no interval (a,b) is empty. Similarly for intervals of the types $[a,b]$, (a,∞) and $(-\infty,a)$.

2. Let $[X,<]$ be a linearly ordered set. If $[X,<]$ is complete-D, then $[x,<]$ is "complete on the left". That is, if M is a non-empty subset of X, and M is bounded below, then M has a greatest lower bound glb M.

3. Every linearly ordered space is Hausdorff.

4. Let M be the unit square $[0,1]^2$ in $\mathbb{R}^2$, and let $<$ be the lexicographic ordering of M. Then $[M,<]$ is complete-D.

5. Suppose that we have given a linearly ordered space $[X,\mathcal{O}]$, where $\mathcal{O} = \mathcal{O}(<)$. Let T be a non-empty subset of X. Then there are two ways to define a topology for T. (1) Let $\mathcal{O}_1 = \mathcal{O}|T$. (2) Let $<_2$ be the restriction of $<$ to T, and let $\mathcal{O}_2 = \mathcal{O}(<_2)$. Question: Do these procedures always give the same topology for T?

6. Let G be a collection of disjoint open sets in a compact Hausdorff space. Then G is countable.

7. In a compact metrizable space, every collection of disjoint open sets is countable.

8. In a compact Hausdorff space, every uncountable set contains a limit point of itself.

9. In a compact metrizable space, every uncountable set contains a limit point of itself.

Chapter 11: Mappings Between Metric Spaces

We recall the definition of continuity, for a function $f: I \to \mathbb{R}$, where I is an interval in $\mathbb{R}$. Let $x_o \in I$, and suppose that for every $\epsilon > 0$ there is a $\delta_{x_o, \epsilon} > 0$ such that

$$x \in I \quad \text{and} \quad |x - x_o| < \delta_{x_o, \epsilon} \implies |f(x) - f(x_o)| < \epsilon .$$

Then f is <u>continuous</u> at x_o. If f is continuous at each point of I, then f is <u>continuous</u>.

We use the notation $\delta_{x_o, \epsilon}$ in order to emphasize that in general, $\delta_{x_o, \epsilon}$ depends not only on ϵ but also on x_o. (Examples?)

Note that $|x - x_o|$ and $|f(x) - f(x_o)|$ are distances in $\mathbb{R}$. In these terms, the definition of continuity can be generalized, so as to apply to functions $f: X \to Y$, where $[X, d]$ and $[Y, d']$ are any metric spaces.

<u>Definition</u>. Let $[X, d]$ and $[Y, d']$ be metric spaces, let A be a set of points in X, and let f be a function $A \to Y$. Let $P_o \in A$. Suppose that for every $\epsilon > 0$ there is a $\delta_{P_o, \epsilon} > 0$ such that

$$P \in A \quad \text{and} \quad d(P, P_o) < \delta_{P_o, \epsilon} \implies d'(f(P), f(P_o)) < \epsilon .$$

Then f is <u>continuous</u> at P_o. If f is continuous at each point of A, then f is <u>continuous</u>. A continuous function is called a <u>mapping</u>.

For the case of functions $f: I \to \mathbb{R}$, this definition agrees with our previous one. Later we shall generalize still further, so as to define continuity for functions $f: A \to Y$, where $[X, \mathcal{O}]$ and $[Y, \mathcal{O}']$ are topological spaces and $A \subset X$.

Let $[X, d]$ be a metric space, let A be a set of points in X, and let f be a function $A \to \mathbb{R}$. Let M and m be real numbers. If $f(x) \leq M$, for each x in A, then M is an <u>upper bound</u> of f. If $m \leq f(x)$, for each x in A, then m is a <u>lower bound</u> of f. A function which has an upper bound (or a lower bound) is <u>bounded above</u> (or <u>bounded below</u>). If f is both bounded above and bounded below, then f is <u>bounded</u>.

Let $x_o \in X$. Suppose that there is a $\delta_{x_o} > 0$ such that $f|N(x_o, \delta_{x_o})$ is bounded. Then f is <u>locally bounded at</u> x_o. If this condition holds for each $x_o \in X$, then f is <u>everywhere locally bounded</u>. (Similarly for <u>locally bounded above at</u> x_o, <u>everywhere locally bounded above</u>, and so on.)

   Theorem 1. Let f be a mapping $X \to \mathbb{R}$. Then f is everywhere locally bounded.

   (Note that this, and many theorems like it, apply automatically to functions $f: A \to \mathbb{R}$, where $A \subset X$; we merely form the subspace $[A, d | A \times A]$.)

   Theorem 2. Let $[X,d]$ be a compact metric space, and let f be a function $X \to \mathbb{R}$. If f is everywhere locally bounded, then f is bounded.

   Theorem 3. Let $[X,d]$ be a compact metric space, and let f be a mapping $X \to \mathbb{R}$. Then f is bounded.

   If f is a function $X \to \mathbb{R}$, and $k \in \mathbb{R}$, then kf is the function under which $x \mapsto kf(x)$. Similarly for the sum $f + g$ of two functions $X \to \mathbb{R}$: $(f+g)(x) = f(x) + g(x)$.

   Theorem 4. If f is a mapping $X \to \mathbb{R}$, and $k \in \mathbb{R}$, then kf is a mapping $X \to \mathbb{R}$.

   Theorem 5. If f and g are mappings $X \to \mathbb{R}$, then $f + g$ is a mapping $X \to \mathbb{R}$.

   A function $f: X \to \mathbb{R}$ is bounded away from 0 if there is an $\epsilon > 0$ such that for each $x \in X$, $|f(x)| \geq \epsilon$. Let $x_0 \in X$. If there is a $\delta_{x_0} > 0$ such that $f|N(x_0, \delta_{x_0})$ is bounded away from 0, then f is locally bounded away from 0 at x_0.

   Theorem 6. Let f be a mapping $X \to \mathbb{R}$, and let $x_0 \in X$. If $f(x_0) \neq 0$, then f is locally bounded away from 0 at x_0.

   Theorem 7. Let F be a mapping $X \to \mathbb{R}$. If X is compact, and f is everywhere locally bounded away from 0, then f is bounded away from 0.

   Theorem 8. Let f be a mapping $X \to \mathbb{R}$. If X is compact, and $f(x) \neq 0$ for each x, then f is bounded away from 0.

   Theorem 9. Let f be a mapping $X \to \mathbb{R}$, such that $f(x) \neq 0$ for each x. For each x, let $g(x) = 1/f(x)$. Then g is a mapping.

   Theorem 10. Let f be a mapping $X \to \mathbb{R}$. If X is compact, then f has a maximum value and a minimum value.

   Note that if the elementary theory of mappings $[a,b] \to \mathbb{R}$ is known, then very little of the above is new; the elementary proofs are readily adaptable to the general case. (Of course, the resulting theorems apply automatically to functions $f: A \to \mathbb{R}$, where $A \subset \mathbb{R}^n$.) The generalization of uniform continuity is equally straightforward, as we shall now see.

   Definition. Let $[X,d]$ and $[Y,d']$ be metric spaces, and let f be a mapping $X \to Y$. Suppose that for every $\epsilon > 0$ there is a $\delta > 0$ such that

$$d(x,x') < \delta \implies d'(f(x),f(x')) < \epsilon .$$

Then f is uniformly continuous.

11. Mappings Between Metric Spaces.

Theorem 11. Let f be a mapping $X \longrightarrow Y$. If X is compact, then f is uniformly continuous.

Problem Set 11.

Prove or disprove:

1. **Theorem.** Let f and g be functions $X \longrightarrow \mathbb{R}$, and let $x_o \in X$. If (1) $f(x_o) = 0$, (2) f is continuous at x_o, and (3) g is locally bounded at x_o, then (4) fg is continuous at x_o.

2. If f and g are mappings $X \longrightarrow \mathbb{R}$, then fg is a mapping.

3. Let f be a function $[0,1] \longrightarrow \mathbb{R}$. Suppose that for each $x_o \in [0,1]$, $\lim_{x \to x_o} f(x) = 0$. Then $f(a) = 0$ for some a.

4. Let f be as in Problem 3, and let $A = \{x \mid f(x) \neq 0\}$. Then A is countable.

5. Let f be a mapping $X \longrightarrow \mathbb{R}$. If P_o, $P_1 \in X$, and $f(P_o) < k < f(P_1)$, then $k = f(P)$ for some P.

6. Let f be a mapping $[a,b] \longrightarrow \mathbb{R}$, and let M be the graph of f; that is, $M = \{(x,y) \mid x \in [a,b] \text{ and } y = f(x)\} \subset \mathbb{R}^2$. Then M is compact.

7. Under the conditions of Problem 6, let $F: [a,b] \longrightarrow \mathbb{R}^2$ be the function $x \longmapsto (x, f(x))$. Then F is a surjective mapping $[a,b] \longrightarrow M$.

8. Under the same conditions, F is a bijection, and F^{-1} is a mapping.

Chapter 12: Mappings Between Topological Spaces

We shall now generalize the definition of a mapping, in such a way that it will apply to functions $f: X \longrightarrow Y$, where $[X,\mathcal{O}]$ and $[Y,\mathcal{O}']$ are any topological spaces. The idea that is needed here is brought out in the following two theorems.

Theorem 1. Let $[X,d]$ and $[Y,d']$ be metric spaces, and let f be a mapping $X \rightarrow Y$. If $V \subset Y$, and V is open (in Y), then $f^{-1}(V)$ is open (in X).

(We recall, from Section 1, that $f^{-1}(V) = \{x \mid f(x) \in V\}$. Thus $f^{-1}(V)$ may easily be empty, and the use of the notation f^{-1} does not mean that f has an inverse.)

Theorem 2. Let $[X,d]$ and $[Y,d']$ be metric spaces, and let f be a function $X \rightarrow Y$, such that if V is open in Y, then the set $f^{-1}(V)$ is open in X. Then f is a mapping.

Definition. Let $[X,\mathcal{O}]$ and $[Y,\mathcal{O}']$ be topological spaces, and let f be a function $X \rightarrow Y$, such that for each open set $V \subset Y$, the set $f^{-1}(V)$ is open in X. Then f is a mapping.

(Query: How should this definition be generalized, so as to apply to functions $f: A \rightarrow Y$, where A is any non-empty subset of X?)

Theorem 3. Let f be a function $X \rightarrow Y$. Then (1) f is a mapping if and only if (2) for each $M \subset X$, $f(\overline{M}) \subset \overline{f(M)}$.

Theorem 4. Compactness is preserved by (surjective) mappings. That is, if $[X,\mathcal{O}]$ and $[Y,\mathcal{O}']$ are topological spaces, and X is compact, and $f:X \rightarrow Y$ is a mapping, then $f(X)$ is a compact set in Y.

Theorem 5. For Hausdorff spaces, BW-compactness is preserved by (surjective) mappings. That is, if $[X,\mathcal{O}]$ is Hausdorff and BW-compact, and $f: X \rightarrow Y$ is a mapping, then $f(X)$ is a BW-compact set in Y.

Let $[X,\mathcal{O}]$ and $[Y,\mathcal{O}']$ be topological spaces, and let f be a bijection $X \leftrightarrow Y$. If both f and f^{-1} are mappings, then f is a homeomorphism of X onto Y. Under these conditions, we have

$$U \in \mathcal{O} \Longrightarrow f(U) \in \mathcal{O}' \ .$$

If there is a homeomorphism $X \leftrightarrow Y$, then the spaces $[X,\mathcal{O}]$ and $[Y,\mathcal{O}']$ are homeomorphic.

12. Mappings Between Topological Spaces

Problem Set 12

Investigate the following.

1. In Theorem 5, did we need to require that $[X,\mathcal{O}]$ be Hausdorff?

2. Theorem (?). Let $[X,\mathcal{O}]$ and $[Y,\mathcal{O}']$ be topological spaces, and let $f: X \twoheadrightarrow Y$ be a surjective mapping. If $[X,\mathcal{O}]$ is Hausdorff, then so also is $[Y,\mathcal{O}']$.

3. Under the conditions of Problem 2, if $[X,\mathcal{O}]$ is metrizable, then so also is $[Y,\mathcal{O}']$.

4. Let $f: X \leftrightarrow Y$ be a bijection. If f is a mapping, then so also is f^{-1}, and f is a homeomorphism.

Following are descriptions of pairs of topological spaces. Try to find out, in each case, whether the two spaces are homeomorphic. In each case, unless the topology is explicitly described, then the "usual" topology is meant. E.g., $\mathbb{R}$ and $\mathbb{R}^n$ have the topology induced by the usual distance, and subsets of $\mathbb{R}$ and $\mathbb{R}^n$ have the subspace topology, unless the contrary is stated.

5. (a) $\mathbb{R}$. (b) A closed interval $[a,b]$.

6. (a) $\mathbb{R}$. (b) An open interval (a,b).

7. (a) a closed interval $[a,b]$. (b) The union of two disjoint closed intervals $[a,b]$ and $[c,d]$.

8. (a) $X = H^+ \cup A$, where H^+ is the upper half-plane $\{(x,y)\,|\,y > 0\}$ and A is the x-axis. (Subspace topology.) (b) $Y = X$. In (b), neighborhoods are of two types. (1) The interior of a circle lying in H^+ is a neighborhood. (2) If C is a circle with center in H^+, tangent to A at P, and U is the interior of C, then $U \cup \{P\}$ is a neighborhood.

9. (a) $X = H^+ \cup A$, with the subspace topology, as in 8 (a). (b) $Y = X$. Neighborhoods are of type (1), as in Problem 8(b), and (2') sets of the form $(U \cap H^+) \cup \{P\}$, where $P \in X$ and U is the interior of a circle with center at P.

10. (a) $Q_0 = Q \cap [0,1]$. (b) $Q_0 \times Q_0$. (Subspace topology in each case.)

11. (a) Q_0. (b) $Q_0 \times Q_0$, with the lexicographic order topology.

12. (a) Any Cantor set on $[0,1]$. (b) Any other Cantor set on $[0,1]$.

13. (a) Any Cantor set C on $[0,1]$. (b) $C \times C$ (with the subspace topology.)

14. (a) Any Cantor set C on $[0,1]$. (b) $C \times C$, (with the lexicographic topology.)

Chapter 13: Connectivity

Roughly speaking, a space X is connected if it is "all in one piece". Thus if
X is a closed interval in ℝ, then X is connected, but if Y is the union of
two disjoint closed intervals, then Y is not connected. This idea is the basis
of the following definition.

   Definition. A topological space [X,𝒪] is <u>connected</u> if X is not the union of
any two disjoint non-empty open sets. A set M ⊂ X is <u>connected</u> if it forms a
connected subspace, that is, if [M,𝒪|M] is connected.

   Often it is convenient to discuss the connectivity of sets M ⊂ X without
introducing subspaces. For this purpose we need the following ideas. Two sets H
and K are <u>separated</u> if H ∩ K̄ = H̄ ∩ K = ∅. (Thus neither of the sets H and K
contains either a point or a limit point of the other.)

   Theorem 1. Given M ⊂ X. Then (1) M is connected if and only if (2) M is
not the union of any two non-empty separated sets.

   It is now obvious that if Y is the union of two disjoint closed intervals in ℝ,
then Y is not connected. But the following requires a proof; it turns out to be
our nth description of the continuity of ℝ.

   Theorem 2. Every closed interval in ℝ is connected.

   Theorem 3. If H and K are separated, and M is a connected set lying in
H ∪ K, then either M ⊂ H are M ⊂ K.

   Theorem 4. Let G be a collection of connected sets, with a point P in
common. Then G* is connected.

   Theorem 5. If M is connected, and M ⊂ L ⊂ M̄, then L is connected.

   Theorem 6. Connectivity is preserved under surjective mappings. That is, if f
is a mapping X ⟶ Y, and M is a connected set in X, then f(M) is connected in
Y.

   Note that Theorem 1 and Theorems 3-6 hold in any topological space; in any
natural schemes of proof, the question of special hypotheses simply does not arise.

   Definition. Let [X,𝒪] be a topological space, and let M be a set of points
in X. A <u>path in M</u> is a mapping p:[a,b] ⟶ M. Suppose that for each P, Q ∈ M
there is a path in M, from P to Q, that is, a path p:[a,b] ⟶ M such that
p(a) = P and p(b) = Q. Then M is <u>pathwise connected</u>.

   Theorem 7. Every pathwise connected set is connected.

   Let M be a set, and let A and B be disjoint sets. If M is the union of
two separated sets, containing M ∩ A and M ∩ B respectively, then A and B are
<u>separated in M</u>. (Here it is not required that A and B be subsets of M.)

13. Connectivity

Theorem 8. In a metric space, let M_1, M_2, ... be a nested sequence of compact sets, and let A and B be disjoint sets. Suppose that for each i, A and B are not separated in M_i. Then A and B are not separated in

$$M_\infty = \bigcap_{i=1} M_i \ .$$

Theorem 9. Let M be a compact set, let A and B be disjoint closed sets, and suppose that A and B are not separated in M. Then there is a compact subset N of M such that (1) A and B are not separated in N and (2) if K is a compact proper subset of N, then A and B are separated in K.

In a metric space, a _continuum_ is a compact connected set.

Theorem 10. In a metric space, let M be a compact set, and let A and B be disjoint closed sets. Then either (1) M contains a continuum which intersects both A and B or (2) A and B are separated in M.

Let M be a continuum, and let P and Q be two points of M. If no proper subcontinuum of M contains both P and Q, then M is _irredicible between P and Q_.

Theorem 11. Let M be a continuum, and let P and Q be two points of M. Then some subcontinuum of M is irreducible between P and Q.

Problem Set 13

Investigate the question of the connectivity of the following sets.

1. $\mathbb{R}$.

2. An open interval (a,b) in $\mathbb{R}$.

3. A half-open interval [a,b) or (a,b].

4. $\mathbb{R}^2$.

5. The graph G of the function $f(x) = \sin(1/x)$ $(0 < x < 1/\pi)$.

6. The union of G and the points (0,1) and (0,-1).

7. The union of G and the linear interval from (0,-1) to (0,1).

8. The space described in Problem 8 (b) of Section 11.

9. The space described in Problem 9 (b) of Section 11.

10. Let x_1, x_2, ... be a sequence in which each rational number between 0 and 1 appears exactly once. For each i, let $\theta_i = 2\pi x_i$, and let K_i be the closed linear interval between the points whose polar coordinates are $(1/i, \theta_i)$ and $(1, \theta_i)$. Let M be the union of all the sets K_i, together with the origin (0,0). Is M connected?

11. $\mathbb{Q}^2$ $(= \mathbb{Q} \times \mathbb{Q} \subset \mathbb{R}^2)$.

12. $M = \mathbb{R}^2 - (\mathbb{Q} \times \mathbb{Q})$.

13. $L = (\mathbb{R} \times \mathbb{Q}) \cup (\mathbb{Q} \times \mathbb{R})$.

14. Let G be as in Problem 5, and let M be the union of G and the origin. Is G homeomorphic to the interval $[0, 1/\pi]$?

15. Let M be as in Problem 12. Is M homeomorphic to $\mathbb{R}^2$?

16. Suppose that in Theorems 1-6, we replace the term <u>connected</u> by the term <u>pathwise connected</u>. Which (if any) of the resulting propositions are true?

In a metric space, a <u>chain</u> is a finite sequence C: c_1, c_2, $\ldots$, c_k of open sets, such that $\bar{c}_i \cap \bar{c}_j \neq \emptyset$ if and only if i and j are consecutive. C^* is the union of the sets c_i. The sets c_i are called the <u>links</u> of C. The <u>end-links</u> are c_1 and c_k. We suppose that each set $\bar{c}_i$ is compact, so that c_i is bounded. The <u>mesh</u> δC of C is the largest of the diameters of the links c_i. Note that the links of a chain are not required to be connected.

17. Let C_1, C_2, $\ldots$ be a sequence of chains, and let P and Q be points of C_1^*. Suppose that (1) for each i, each link of C_{i+1} lies in some link of C_i (so that $C_{i+1}^* \subset C_i^*$), (2) $\text{Lim}_{i \to \infty} \delta C_i = 0$, and (3) for each i, P and Q lie in the end-links of C_i. Let $M = \bigcap_{i=1} \bar{C}_i^*$. Then M is a continuum, and is irreducible between P and Q.

An <u>arc</u> is a set which is homeomorphic to a closed linear interval $[a,b]$. If there is a homeomorphism $f: [a,b] \leftrightarrow M$, $a \mapsto P$, $b \mapsto Q$, then P and Q are <u>end-points</u> of M.

18. Under the conditions of Problem 17, M is an arc, with end-points P and Q.

19. No continuum M contains three points such that M is irreducible between every two of them.

20. Every arc has two and only two end-points. (This requires a proof, because if there is one homeomorphism $f: [a,b] \leftrightarrow M$, then there are plenty of others.)

Chapter 14: Well-ordering

The following is familiar.

<u>The Well-ordering Principle</u>. Every non-empty subset of $\mathbb{Z}^+$ has a least element.

More generally, a <u>well-ordered sequence</u> (WOS) is a linearly ordered set $[S,<]$ in which every non-empty subset has a least element. The elements of S are called the <u>terms</u> of the WOS. The relation $<$ is called a <u>well-ordering</u> of S. Ordinarily we denote a WOS by a single Greek letter, and we write

$$\alpha: a_0, a_1, \ldots,$$

where a_0 and a_1 are the first two terms. α^* is the set of all terms of α. In a way, the use of integers as subscripts is misleading, because α may be much more complicated than $[\mathbb{Z}^+,<]$. For example:

<u>Theorem 1</u>. Let α and β be well-ordered sequences. Then the lexicographic product $\alpha \times \beta$ of α and β is a WOS.

(The terms of the product are the elements (a,b) of $\alpha^* \times \beta^*$; and $(a,b) < (a',b')$ means, by definition, that (1) $a < a'$ or (2) $a = a'$ and $b < b'$.)

The Well-ordering Principle says that in $[\mathbb{R},<]$, the ordered set $[\mathbb{Z}^+,<|\mathbb{Z}^+]$ is a WOS. Of course $\mathbb{Z}^+$ is countable. And this is what always happens:

<u>Theorem 2</u>. If $M \subseteq \mathbb{R}$, and $<|M$ is a well-ordering of M, then M is countable.

Let $[A,<]$ and $[B,<']$ be ordered sets, and let $f:A \longleftrightarrow B$ be a bijection. If $x < y \iff f(x) <' f(y)$, then f is an <u>order-isomorphism</u>. If there is such a bijection between A and B, then $[A,<]$ and $[B,<']$ are <u>isomorphic</u>.

In these terms, we have a sort of converse of Theorem 2, as follows.

<u>Theorem 3</u>. Let α be a WOS, and suppose that α^* is countable. Then there is a subset M of $\mathbb{R}$ such that α and $[M,<|M]$ are isomorphic.

Let α be a WOS, and let B be a subset of α^*. Then $[B,<|B]$ is a WOS β. Suppose that for each $b \in B$, $a < b$ in $\alpha \Rightarrow a \in B$. Then β is called a <u>segment</u> of α, and we write

$$\beta \leq \alpha .$$

In this case, we say that B <u>forms</u> a segment of α. If $\beta \leq \alpha$ and $\beta \neq \alpha$, then β is a <u>proper segment</u> of α, and we write

$$\beta < \alpha.$$

Evidently every WOS α is a segment of itself. And if $a \in \alpha^*$, and

$$B = \{b | b < a \text{ in } \alpha\} ,$$

then the corresponding WOS β is a segment, because $<$ is transitive. This segment is denoted by α_a. Similarly, the set

$$B' = \{b|b \leq a \text{ in } \alpha\}$$

forms a segment, with a as its last term. This segment will be denoted by α_{a+}.

Theorem 4. $\beta < \alpha$ if and only if $\beta = \alpha_a$ for some $a \in \alpha^*$.

Theorem 5. Let α be a WOS, and let G be a collection of subsets of α^*, each of which forms a segment of α. Then (1) the union of the elements of G forms a segment of α and (2) the intersection of the elements of G forms a segment of α.

Theorem 6. Let $f: \alpha^* \longleftrightarrow \beta^*$ be an isomorphism between the WOS α and β. If A forms a segment of α, then $f(A)$ forms a segment of β.

Theorem 7. Let α and β be WOS. Then one of them is isomorphic to a segment of the other. And the isomorphism f, between α and a segment of β, or between β and a segment of α, is uniquely determined by α and β.

This says, in effect, that two WOS can differ only in length.

Theorem 8. Let α and β be WOS. If $\beta \leq \alpha$, and α and β are isomorphic, then $\alpha = \beta$. In fact, if (1) $B \subset \alpha^*$, (2) B forms a segment of α, and (3) f is an isomorphism $\alpha^* \longleftrightarrow B = \beta^*$, then (4) f is the identity.

Problem Set 14

1. Suppose that $[S,>]$ is a well-ordering. Define $a <' b$ to mean that $b < a$. This gives an ordered set $[S,<']$. If $<'$ is also a well-ordering, what can you conclude about S?

2. Suppose, throughout this problem set, that there is a WOS α such that α^* is uncountable. (This is true, as we shall see in the following section.) Show that there must be a WOS β such that (1) β^* is uncountable, but (2) for each term b of β, β_b^* is countable. Such a WOS is called a sequence of type Ω.

3. If α and β are sequences of type Ω, then α and β are isomorphic.

4. Every infinite set contains a countably infinite set.

5. Let α be a WOS of type Ω. Consider the topological space $[\alpha^*, \mathcal{O}(<)]$.

    (a) Is it metrizable?

    (b) Is it separable?

    (c) Is it BW-compact?

    (d) Is it compact?

6. Let α be a WOS, and suppose that α is rigid, in the sense that the identity is the only isomorphism of α into α. What can you conclude?

7. Let $[S,<]$ be $[0,1]^2$, with the lexicographic order. Let A be a subset of S, and suppose that $[A,<|A]$ is a WOS α. Does it follow that A is countable?

Chapter 15: The Existence of Well-orderings. Zorn's Lemma.

The following is analogous to the Unique Finite Sequences Theorem of Section 9.

Theorem 1. (The Unique Transfinite Sequences Theorem.) Let A be a collection of well-ordered sequences, such that if $\alpha, \beta \in A$, then one of the sequences α and β is a segment of the other. Then there is a WOS δ such that for every element of A is a segment of δ.

Theorem 2. Let S be any set, let G be a collection of proper subsets of S, including the empty set $\emptyset$, and let

$$\emptyset: G \longrightarrow S$$

be a function such that for each $g \in G$,

$$\emptyset(g) \in S - g.$$

Then there is a WOS

$$\alpha = a_0, a_1, \ldots$$

such that (1) $a_0 = \emptyset(\phi)$, (2) if $a \in \alpha^*$ and $a \neq a_0$, then $a = \emptyset(a_a^*)$, and (3) $\alpha^* \not\subset G$.

The statement of this theorem is admittedly a mouthful; but once the theorem is stated, it is easy to prove and to use. It has a corollary which in many cases is harder to apply.

Theorem 3. (The Well-ordering Theorem; E. Zermelo.) Every set can be well-ordered. That is, for every non-empty set A there is a WOS α such that $\alpha^* = A$.

Let $[S,<]$ be an ordered set. A subset M of S is <u>linearly ordered</u> if $<|M$ is a linear ordering of M. An element b of S is an <u>upper bound</u> of a set $A \subset S$ if $a \leq b$ for each a in A. An element m of S is <u>maximal</u> if $m < a$ is false for each a in S.

(<u>Warning</u>: Slight variations on these definitions turn out to be "wrong" definitions; they give wrong answers. For example, if an upper bound of a set A is required to be strictly greater than each element of A, then it turns out that $S = [0,1]$, with the usual ordering, has no upper bound. The same sort of trouble arises if we require that for m to be a maximal element, we must have $m \geq a$ for each a in S. For example, we may have $S = [0,1]^2$, with $(x,y) < (x',y') \Longleftrightarrow y < y'$. All the points $(x,1)$ are maximal, under the "right" definition, but none of them are maximal under the "wrong" definition.)

Theorem 4. (Zorn's Lemma.) Let $[S,<]$ be an ordered set, and suppose that every linearly ordered subset of S has an upper bound. Then S has a maximal element.

Topology

For most purposes but not quite all, this supersedes the complications of
Theorem 2.

Problem Set 15

1. Every infinite set contains a countably infinite set.

2. Let A and B be any non-empty sets. Then one of them is cardinally
equivalent to a subset of the other.

3. Show that the Axiom of Choice is a consequence of Zorn's Lemma.

4. Use the Well-ordering Theorem to get a very quick proof of the Choice of
Function Theorem.

5. Let A be any non-empty set. Then A has a minimal well-ordering. That
is, there is a well-ordering α of A such that no proper segment of α is isomorphic to any well-ordering of A.

6. Use well-ordering to get an easy proof of the Schröder-Bernstein Theorem.
(Such a scheme was used by Georg Cantor, to get the first proof of the theorem.)

7. Let [X,d] be a metric space in which every uncountable set has a limit
point. Show that for each $\epsilon > 0$ there is a countable set M such that $N(M,\epsilon) = X$.

8. Let [X,d] be as in Problem 7. Then [X,d] is separable (and therefore
completely separable).

Universitext

Editors: F.W. Gehring, P.R. Halmos, C.C. Moore

Universitext

Editors: F.W. Gehring, P.R. Halmos, C.C. Moore

Chern: Complex Manifolds Without Potential Theory
Chorin/Marsden: A Mathematical Introduction to Fluid Mechanics
Cohn: A Classical Invitation to Algebraic Numbers and Class Fields
Curtis: Matrix Groups
van Dalen: Logic and Structure
Devlin: Fundamentals of Contemporary Set Theory
Edwards: A Formal Background to Mathematics I a/b: Logic, Sets and
    Numbers
Edwards: A Formal Background to Mathematics 2. A Critical Approach to
    Elementary Analysis
Frauenthal: Mathematical Modeling in Epidemiology
Fuller: FORTRAN Programming: A Supplement for Calculus Courses
Gardiner: A First Course in Group Theory
Greub: Multilinear Algebra
Hajek/Havránek: Mechanizing Hypothesis Formation: Mathematical
    Foundations for a General Theory
Hermes: Introduction to Mathematical Logic
Kuffrecht: Probability and Statistics at Historical Pi
Kelly/Matthews: The Non-Euclidean Hyperbolic Plane: Its Structure and
    Consistency
Kostrikin: Introduction to Algebra
Luecking: Singularity Theory and an Introduction to Catastrophe Theory
Marcus: Number Fields
Naber: Essential Mathematics for Applied Fields
Moise: Introductory Problem Courses in Analysis and Topology
Oden/Reddy: Variational Methods in Theoretical Mechanics
Reisel: Elementary Theory of Metric Spaces: A Course in Constructing
    Mathematical Proofs
Rickart: Natural Function Algebras
Schreiber: Differential Forms: A Heuristic Introduction